The Massive MIMO Report

Massive MIMO Report

Wireless Technology Analyst Update

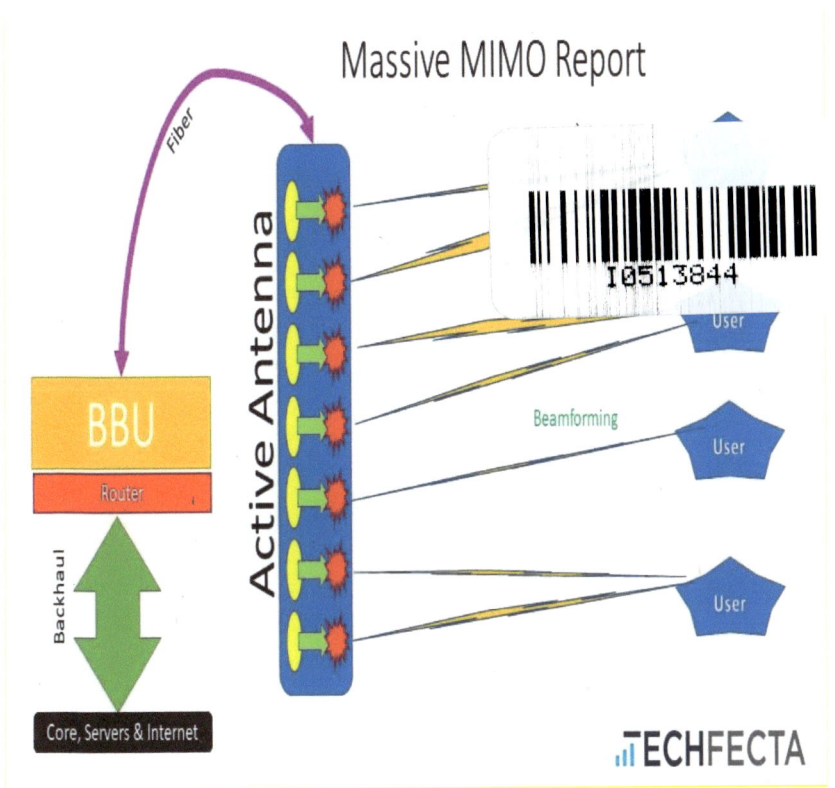

Figure 1 Cover

Massive MIMO Report

Contents

Massive MIMO Report 2
Copyright 7
Thank you! 7
Overview: 8
Introduction: 10
What is MIMO? 10
    Where did MIMO come from? 13
    Are there different types of MIMO? 15
What is massive MIMO? 16
    Why do we need massive MIMO? 17
    Will massive MIMO be needed everywhere? 19
    What Parts make up Massive MIMO in the system? 20
    What does the BBU need to do to support massive MIMO? 21
    What does the Active Antenna System have to do? 22
What makes massive MIMO special? 23
What is beamforming? 25
    First, a quick, high level, history lesson 25
    How does it work? 26
    What spectrum does beamforming work in? 28
    Who will use it? (Looking at the USA only) 28
    Why cable companies should pay attention. 30
    Learn more: 31
The network matters! 32
    What about the backhaul? 32

- What about the fronthaul? 34
- How will the network meet the demands of 5G? 35
- What about the extensions of the macro sites? 35
  - Will CRAN or C-RAN be a massive MIMO system? 35
    - Resources: 37
  - Will Small Cells have Massive MIMO 38
  - What about DAS systems? 38
- Will Massive MIMO be in the UE device? 40
- What changes will tower companies see at the site? 42
  - On the Tower: 43
  - On the ground: 44
- Will Utility costs change? 44
- Massive MIMO Tower Work Overview 46
  - Overview 46
  - What is Massive MIMO, really? 46
  - What about the tower work? 47
    - What if you swap? 48
    - What is it's new? 49
    - Who decides what mount is safe for massive MIMO antennas? ... 49
    - What about the cables? 50
    - Is it bigger or smaller? Size and weight matter! 50
  - How Will TIA-222 Rev H affect Massive MIMO Tower Work? 51
    - What is TIA-222? 51
    - Why does Rev H matter? 51
    - Why now? 52
    - How will this affect new deployments? 52
    - How does this impact 5G? 53

    Resources for TIA-222: 53

        Summary 54

    Tower Crew Summary: 56

        Resources: 56

What does it mean for the suppliers and GCs? 58

    Who benefits? 58

    Who doesn't benefit? 60

Economies of size with Massive MIMO 62

    Overview 62

    Why does size matter? 62

    We'll look at what effects the size. 63

    What about weight? 63

    Is there a difference between TDD and FDD? 64

        What is FDD? 64

        What is TDD? 64

    What about frequency? 67

    How much is too much? 68

    Larger antennas cost more. 68

    How has this changed from the traditional models? 69

    But wait, that's not the big picture! 70

    Pros and Cons: 71

    Resources: 72

Spectrum Options 74

    Mobility Connections: 74

How will Carriers deploy massive MIMO? 75

New Business Models for the Carriers: 77

    Internet Service Providers 77

- TV and Video ... 77
- IOT ... 78
- Transportation ... 79
- Should 5G be Fixed and Mobile Wireless? ... 80
  - How does this tie into massive MIMO? ... 80
  - What's the difference? ... 80
  - Fixed Wireless Overview ... 81
  - Mobile Overview ... 83
  - Why compare fixed to mobile? ... 84
  - Fixed Pros and Cons ... 84
  - Mobile Pros and Cons ... 85
  - Who wins? ... 86
  - Summary: ... 87
  - If you want to learn more: ... 87
- MIMO Report Summary: ... 89
- More Resources: ... 90
- Acronyms and Definitions ... 92
- Thank you! ... 94
- Copyright ... 94
- About ... 95
- More Reports and Books: ... 95

The Massive MIMO Report

Figure 1 Cover 1
Figure 2 MIMO Models 12
Figure 3 Massive MIMO Models 16
Figure 4 Massive MIMO Antenna 25
Figure 5 Beamforming 1 27
Figure 6 Beamforming 2 29
Figure 7 Beamforming 3 30
Figure 8 CRAN and oDAS 37
Figure 9 Massive MIMO FDD Model 67
Figure 10 Massive MIMO TDD Model 67
Figure 11 Back Cover 97

Copyright

First Edition © 2018 by Wade Sarver. All rights reserved. No part of this publication may be reproduced, stored in a retrieval system, or transmitted in any form or by any means, electronic, mechanical, photocopying, recording, scanning, or otherwise, except as permitted under Sections 107 or 108 of the 1976 United States Copyright Act, without the prior written permission of the author.

I am not a lawyer or an actively certified safety expert. This book is completed based on research and my experiences. Safety processes and procedures are constantly updated and improved over time. The material contained is for reference only and may include products, information, or services by third parties. I do not assume responsibility for any third-party material referenced in this book.

This document is a guide to help people and not a guarantee that you will do everything properly. By reading this, you agree that myself and my company is not responsible for the success or failure of your business decisions relating to the information presented in this guide.

www.wade4wireless.com

www.techfecta.com

Cover and design by Wade Sarver

Thank you!

Thank you for purchasing this report, I appreciate your support. I pray that it serves you well.

If you can, let me know what you like and didn't like about this report. What should I add, and what should I skip next time.

If you need one on one consulting or specific reports, feel free to reach out at wade@techfecta.com for direct support.

Overview:

The idea of massive MIMO is revolutionary. The fact it's going to market is exciting. We all looked at 5G, then overlooked how we were going to get there, solve the problems that 5G is introducing. Here is an overview of massive MIMO and what to expect now that all the smart people have made it a reality.

I have a blog that I pulled a majority of this information from. It served as the foundation for what you're reading here.

The report starts out with explanations of MIMO then massive MIMO so that you understand what it is and how it will work. It also gives an understanding of why it's being deployed.

Then the report covers beamforming. This had to be broken out and explained so that you understood what makes massive MIMO so special. This is an extreme kind of beamforming that will revolutionize the way macro sites communicate with the UE device. The end user wants more bandwidth, but that's not what they really want. They want steady and reliable bandwidth to their device. Massive MIMO enables that to happen.

Then, we cover the network. What good is a kick-ass wireless link if your backhaul is crap? Not much!

The report gets into more ways to deploy massive MIMO and an overview of what they are and if massive MIMO makes sense. You probably want to consider how what, and where massive MIMO can be deployed. It's a good idea to see if CRAN and small cells will be part of the massive MIMO ecosystem.

The next section is going to look at how deployment on towers and rooftops will be influenced by massive MIMO systems. I get asked all the time about the tower and how tower companies will react. It's not necessarily the tower companies that will be concerned about the new equipment. It's how the equipment is deployed and in what spectrum. How will installers deal with the new equipment? How will site acquisition companies work on this and what new costs will be incurred by rolling out massive MIMO. All of this is covered.

Speaking of deploying, why does size matter and what is the determining factor of size? Well, it's spectrum among other things. Also, TDD or FDD and why it matters. This rolls into the spectrum section which is just an overview of what is out there and yet to come, in the USA at least.

This rolls into new business models for the carriers. The new business models depend on fixed or mobile systems being deployed. Which will be the 5G focus first?

Of course, there is a summary section to help you look at what's important.

The end has resources as well as acronyms and definitions. This will help you figure out what all those acronyms mean that you're forced to deal with. In this industry, there are so many terms and groups of letters that twist your eyes when looking at them. Why are there so many? Because if we actually had to say everything in log form, we would never get anything done. Unfortunately, by the time the letters are used to describe something over a long period of time, we forget what the stand for. For example, LTE. Do you remember that LTE stands for or why it was named that? Maybe you don't care, but it's Long Term Evolution. Why that? Because it was supposed to last a very long time as the format continued to improve. That is until 5G came out and now we say "5G" all the time. Unfortunately, LTE has lost l's charm to most people. Not to me, I think that long-term evolution is such a cool idea, like something that homo sapiens should be doing continuously. We should be evolving on a regular basis. I mean our knowledge, not like mutants or anything.

So, read this if you dare to learn more about massive MIMO!

The Massive MIMO Report

Introduction:

The Massive MIMO report is for information services only. This forms the research at Techfecta and the opinions of the author. The idea is to prepare the industry for the new system rolling out and what to expect.

To make things easy for you, I added Acronyms and Definitions to the end of this article. That may help if you get confused or want clarification.

Notice that I have made a resources section to help you see where I get my information and how I came to the conclusions I have. This way, if you want to, you can research your specific topic with greater depth.

You may wonder if massive MIMO is a step closer to 5G. It is because it will enable the massive broadband requirement to happen. Not necessarily the traditional way. While it allows more broadband to get to an individual user, the key is that it allows multiple users to get the most broadband simultaneously. That is the key here. Massive MIMO will open up beamforming to multiple specific users simultaneously. OK, that was a lot of big words, so let's look at it this way, it can allow many users to access all the bandwidth at one time. They no longer have to share the time on the antenna or radio head because they can all talk at the same time. It's like going from a party line where everyone had to wait their turn to talk to having 10 dedicated lines to 10 separate people.

First, let's look at MIMO and build the story into massive MIMO.

What is MIMO?

Well, in wireless technology MIMO means that you have several antennas and radios all on one BTS. So, you have one BTS with a sector that has multiple antennas for transmit and receive. So, the alpha sector could have four transmit and four receive all in one panel. With only a few transmit and receive, the radio head could be behind the antenna. MIMO is already deployed across the US by most all the carriers helping LTE reach the speeds it has today. That along with carrier aggregation, improved processing, optimization techniques, and a myriad of additional features making that smartphone look faster than your laptop when running an app.

To put MIMO in perspective, older antennas were SISO, Single in and Single Out; it's just we didn't know we could do more. In the old days, you know, back in the 1990s, we began to realize that with QPSK, Quadrature Phase Shift Keying is a type of modulation that has been around for decades. I was using this back when we did paging systems, (you may know them as beepers). That led to QAM which pushed the boundaries of data through the air. Learn more about the technical side at http://iitg.vlab.co.in/?sub=59&brch=163&sim=1065&cnt=2404 and https://www.rcrwireless.com/20160901/test-and-measurement/what-is-64-qam-tag6-tag99 and http://www.yourdictionary.com/256-qam.

Learn more at https://www.maximintegrated.com/en/app-notes/index.mvp/id/686.

I bring up the technical side because the new modulation opened doors to new ways to receive the signal which built a need for new antennas that created an opportunity for smart antennas. It's an evolution that most people don't care about, but this is how things progress, one breakthrough at a time.

The change in antenna technology to allow more elements to broadcast simultaneously opened new doors for wireless communication. While this concept is not new, the microwave systems were able to do diversity and sharing for years; it is a breakthrough for mobility. Now that the device itself can have 2 to 4 antennas in a small unit all working together has increased throughput to all devices. It's been a game changer.

The maximum throughput race has been on, but the OEMs have not been able to pass anyone, they can only wait for the standards to be released and the technology to catch up. Then the infrastructure OEMs work with the chip manufacturers to get something out that can be tested. Then they must get the smartphone device makers to implement what they have done and test the system end to end. While this sounds fun, it's not always easy. The chipmakers find problems, the infrastructure OEMs have problems, and the devices are hard to put together for a real-world test. Put all of that in with a carrier who is willing to open the cost and expense of a lab or a live site to test, and it's an expensive venture. One that has an enormous payback, if it works. The carrier and OEMs all get to brag about what they did and then all they must do it wait, 6 months or so until it goes from the lab to the real world. While 6 months sounds slow, it is quite fast. There is a lot of testing, approvals, regulations, deployments, inventory, and other issues that all cause delays in the release of new features. It's more than turning it on and watching it work. It's a process, and even if the infrastructure can do it, the devices need to have the feature. If it's a hardware upgrade, then the design of the device matters.

No easy task, but one that we all want to see. Evolution at its finest.

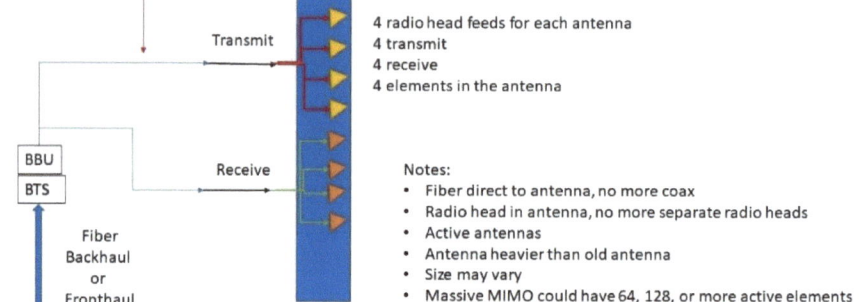

Figure 2 MIMO Models

Where did MIMO come from?

According to Wikipedia, https://en.wikipedia.org/wiki/MIMO, MIMO was developed by Jack Salz of Bell Labs in the mid-1980s. He looked at multi-channeled digital transmissions on wire pairs and how the crosstalk between the wires in bundled pairs had to be dealt with. Good old Jack thought about using the same idea for "mutually cross-coupled linear networks with additive noise sources" according to Wikipedia. He took that knowledge and applied it do dual polarized radio systems. WOW!

This method was brought to cell systems in the 1990s. The idea was to be able to reuse spectrum more efficiently. They had Space-division Multiple Access (SDMA) directional antennas using this. Personally, I saw this in the point-to-multipoint communications before cellular picked it up.

Then, in 1991, ArrayComm researchers Richard Roy and Bjorn Ottersten got the US patent for this method. The idea was that they could get an array of receiving antennas on one base station.

Then it went crazy from there. Researchers kept improving it, like MIMO-OFDM. Cisco got into the act later with their MIMO system. Bell Labs continued to improve the idea. So many companies had concepts, not many could release it commercially. The business case was not always there.

Luckily, we had wireless LANs using this technology. There was the use of it in 3G. It was 4G where we saw MIMO take off to get more data throughput to the end user.

Then we had WiMAX where this technology could be used in the real world. Clearwire, the Sprint-owned company, brought it to the national level when they build out their system. It was a 2.5GHz TDD WiMAX system. The first of its kind. Unfortunately, it didn't last because all the other carriers, including Sprint, were moving towards LTE which had better qualities for mobility.

Don't you worry, LTE systems picked this up and ran with it. The carriers knew this was a better way to go. It cost more at the cell site because now they had to install more expensive antennas and multiple radio heads. But it worked, 4x4 MIMO and 8x8 became common at the tower sites.

Are there different types of MIMO?

The short answer is yes. There are several different types of MIMO.

- **SU-MIMO** – This is single-user MIMO. This is MIMO, but dedicated to a single user, hence the name single-user MIMO. This is great if you only want to connect one device and get awesome bandwidth. However, real-world users like it when more than one device is connected to the BTS, whether it's Wi-Fi or LTE, it's nice when more than one user gets the benefits of MIMO. SU-MIMO allows more throughput to go be passed between the device and the source by having multiple radiating beams communicate information simultaneously. OFDM allowed this to happen and was used in the earlier days for error correction.
- **Macro diversity** – this was a form of MIMO that helped the end user get the best signal to the device by choosing the best signal with multiple antennas. It used a single source but was using multiple receive antennas. Close to MIMO.
- **CO-MIMO** – is Cooperative MIMO which has morphed into MU-MIMO, sort of. This was the stepping stone that allowed MIMO to go beyond one user to many users. I thought it might help you to see the transition to MU-MIMO. Cooperative MIMO was a way for multiple beams to communicate from different base stations to a user simultaneously. Each beam would carry individual carrier information. This would allow the end user to communicate with different sources at the same time, more information to the end user.
- **MU-MIMO** – this is multi-user MIMO, what the hell is that? Well, I am glad you asked. This is where MIMO is pushing more data but to multiple users. Another great thing is that a device that doesn't have MIMO will see the benefits because the BTS with MIMO can communicate with that device specifically and others at the same time. Smart!
- **Massive MIMO** – is the MIMO that has concentrated radio heads that can work together to serve multiple users. It combines multi-user MIMO with beamforming. Generally, it is an array of radiating elements concentrated 16 or more horizontal by 16 or more vertical.

What is massive MIMO?

This is where we have more active elements in the antenna; this could be 32T32R, 64T64R or 128T128R or even higher. The 64T is 64 transmit elements, and 64R is 64 receive elements. Just imagine the feature we get from MIMO, listed above, are suddenly on steroids and making the current features even better! WOW!

This changes many things on the tower. Since running 64 coax jumpers is not practical, the OEMs are inclined to create an active antenna. That would be an antenna with the radio head inside of the antenna so that they could have a connection between each radio head to each element. It just makes sense. No more coax and fiber direct to the antenna. No more radio head and power to the antenna.

What changes will happen in the eNodeB? The BBU will need to handle more processing than ever! They will also need to improve the BBU, the baseband unit.

The key to massive MIMO being successful is the MU-MIMO, multi-user MIMO so that multiple users can access the bandwidth without worry of the next user taking too much of it. The active antenna helps to make this happen. It takes a slew of radio heads and antenna elements to make this happen.

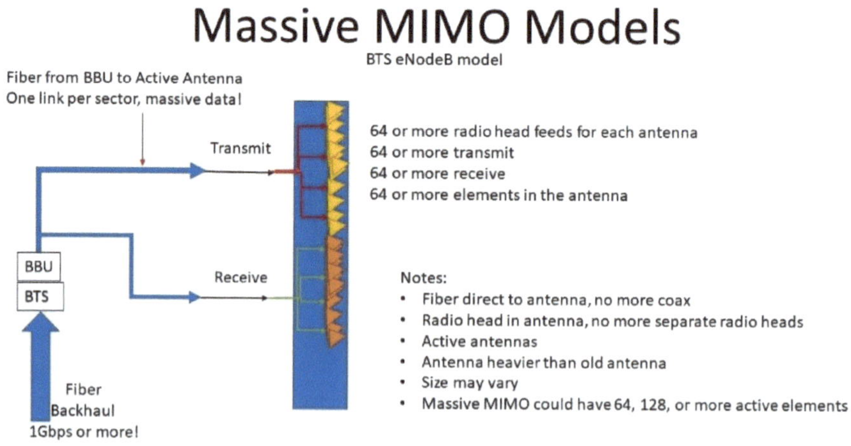

Massive MIMO Models
BTS eNodeB model

Fiber from BBU to Active Antenna
One link per sector, massive data!

Transmit

Receive

BBU
BTS

Fiber Backhaul
1Gbps or more!

64 or more radio head feeds for each antenna
64 or more transmit
64 or more receive
64 or more elements in the antenna

Notes:
- Fiber direct to antenna, no more coax
- Radio head in antenna, no more separate radio heads
- Active antennas
- Antenna heavier than old antenna
- Size may vary
- Massive MIMO could have 64, 128, or more active elements

Figure 3 Massive MIMO Models

Why do we need massive MIMO?

When you look at the evolution of massive broadband and 5G, you will need massive MIMO. Remember the promise of massive broadband everywhere? That is going to be achieved using several tools in the carriers and OEMs arsenal. It will include carrier aggregation and massive MIMO in the urban areas where the need is the greatest.

The urban areas need to have more and more devices connected, simultaneously, while providing large broadband pipes. This is going to solve that problem. It's going to provide the broadband to the end user, or it could be the backhaul to the small cells as required.

The key here is to have a physical upgrade to the site. The carriers hope this is one of the last physical upgrades to the BTS, at least the macro sites. They are going to get LTE to go as fast as it can so that when they say it's 5G, it acts like 5G.

I know most people think that LTE will not be 5G, but the carriers are looking at LTE as being the foundation for 5G because they don't want to do a complete overhaul of the entire system to get it to 5g, hence LTE which stands for Long Term Evolution.

So, we need to get to massive MIMO not only to provide spectacular bandwidth to the users but to have all the parts in place to create a true 5G site. This way the carriers have a 5G ready station at the site.

Massive MIMO really is the stepping stone to 5G, not only to increase bandwidth and the user experience but to get a macro base station in a place that only requires a firmware upgrade to make the leap to 5G. It's critical to put the building blocks in place to maximize the services at a site.

Not only that, but the increased functionality is also part of newer architecture which fits into the same if the not smaller physical footprint of the previous hardware. That means that the OEMs are figuring out how to go small so that the carriers pay less at the tower site. While this is "getting there," it's a step in the right direction to merge the radio heads and the antennas into a smaller and somewhat lighter form factor. Why? Because the carriers want to start reducing OpEx and the easiest way to do that is to pay less for rent at the tower site. The easiest way to pay less is to have a smaller footprint on the tower and ground which should reduce rent.

In 2017 the carriers made it clear that they are tired of paying high rent at tower sites. While this is a valid complaint, they are locked in for several reasons. The coverage is important; they can't start removing sites and lose coverage. The leases are almost ironclad, most carriers are locked in for over 5 years at every site, and the tower companies love this. The equipment has been getting bigger, and with each new spectrum auction, the carriers were adding radio heads and bigger antennas or more antennas. This is changing now as the carriers are pressuring the OEMs to cram more spectrum and power and carriers into one radio head. Massive MIMO may help that if they can cram it all into a radio head and antenna. The growth may be difficult, but we'll wait and see how that pans out.

I am not sure how the growth will happen with new spectrum. The larger carriers, Verizon and AT&T, try to be proactive, they like to have broadband radio heads if possible so that they can easily add spectrum. It doesn't always work out. They also may think they have enough spectrum and try to utilize carrier aggregation and existing systems to grow bandwidth. They are doing it already.

The other thing that carriers hope to get from massive MIMO is lower latency. The new equipment is faster than ever. However, the latency will be more than the new hardware in the air. They need to rely on the networking improvements. It will take the system design to improve latency.

Let's sum it up:

- High bandwidth needed for throughput.
- Spectral efficiency for improved spectrum usage.
- Lower latency.
- Improved device density connectivity, densification.
- It is a step towards 5G.
- The smaller form factor on the tower (thank all the radio heads, coax, and antenna.

Will massive MIMO be needed everywhere?

No, in fact, until there is a need, the carriers won't install it. The need in the urban areas is now. They need to get the most out of the macro sites, and massive MIMO will allow carriers to increase bandwidth and UE loading. That is what they need right now, especially in urban area, any city with a dense population will need to be taken care of.

The key will be to balance the payback with the deployment. Urban areas need this technology now. Suburban and rural, not so much. The loading and bandwidth demands are in the cities. If you ask T-Mobile, they will be the first to admit that they concentrate their growth in urban areas, but they do their testing in rural areas. Rural affects fewer people. Why spend the money there until you have too?

Urban areas are already overloaded, and the carriers are not looking at small cells in the same way they did 5 years ago. They see the limitations and know the limitations of the small cell, which happens to be loading and throughput. What? That's why we put small cells in, right? Like the mini macro, we use that do, right? NO! We put them in to offload some of the traffic from the macro site, just like using Wi-Fi which is way cheaper but shared and less efficient. Massive MIMO will eliminate the need for many small cells almost instantaneously. They provide the coverage and densification whereas the small cell could only provide it in one small specific area. The massive MIMO site can provide loading and throughput to larger areas and for more devices per sector. WOW!

Eventually, the massive MIMO systems will be everywhere, but I don't see it happening in a year or 2, maybe more like 5 or 6. Why would the carrier spend the money if they don't see a payback? Carriers are savvier now since LTE and 5G will talk to each other seamlessly. The investment must have some type of payback.

It's not like going from 2G to 3G to 4G since those systems had interoperability issues. Hence, LTE! Long Term Evolution makes the interoperability a requirement for different systems with a seamless handoff with similar features.

What Parts make up Massive MIMO in the system?

When looking at the massive MIMO system, let's look at what is changing at the site. The radio head and antenna will be removed and replaced by an active antenna. This should mean less equipment on the tower, but it may not be lighter. You would hope the new technology is lighter, but all the weight now is in the antenna, not distributed between a radio head and antenna.

The BBU will need to be upgraded to handle the throughput. It's going to take a faster processor to handle talking to 32 or 64 or 128 elements on the radio head simultaneously. It's no easy task and quite demanding. WOW! That a lot of communication going on at one time through one device!

Then we have the fronthaul, the connection between the BBU and the active antenna. That is probably going to need more than 1 or 2 fiber pairs. It may need 10 to 20 to communicate and pass the traffic. It's not just the traffic for all those individual elements, but also the overhead to control each element, communicate with each UE, manage the beam width of each element, the scheduling of each device. Now you see, the traffic demand between the BBU and the active antenna will be huge!

Let's not forget the backhaul demand. The router will need to be upgraded as will the backhaul bandwidth. It needs to be increased exponentially. I don't see 100Mbps handling any of this anymore. While that may be a good place to start, it's probably going to ramp up to 10Gbps just to maintain the traffic at any given urban site throughout the day.

The core will need to be expanded to handle the excessive traffic that the site brings in on each sector, but that is another part of the system.

Let's sum it up:

- The active antenna system
- The BBU upgrade
- The fronthaul, the link between the BBU and the active antenna, may need more than 5 fiber pairs per sector.
- The router needs to be upgraded.

- The backhaul needs more bandwidth, over 10 Gbps at any site, maybe more.
- The core will need to be upgraded to handle the new service and additional traffic at each site.

What does the BBU need to do to support massive MIMO?

It needs to process so much more data. Now, instead of controlling one or 8 radio heads, it will need to process and distribute the power across 64 or more individual radio elements. It needs to pass more data.

The BBU needs to be significantly improved. What will it do?

- Process more data.
- More bandwidth.
- Control more radio heads.
- Talk to more UE devices.
- Handle carrier aggregation.
- Process more services simultaneously.
- Perform self-optimizing network, SON, services.
- Capture data.
- Handle neighbor list to avoid interference.

OK, I said 8 radio heads, but the reality is that the radio head just has 8 to 16 ports on it. It would have one for transmit and one for receive for each element in the antenna. So technically you would think there would be a radio head for each antenna, but the reality is that one radio head will control all the functions and feed one antenna with numerous elements.

Let's face it; the real world must meet the tech world. We need to make sure we don't overload the towers with excessive equipment. The hardware issue has been resolved, and the OEMs found ways to make things smaller and better. Finally, we can have more services on the tower but smaller equipment with less wind loading, and lighter than before.

It also needs to be smart. The BBU finds it's easier to talk to a smart radio head, just one, rather than 64 at the same time. It helps to have control of the local site so that the signals don't interfere with each other nor the neighboring sectors and sites.

The equipment is beginning to get smaller and smarter, more on that in this report.

What does the Active Antenna System have to do?

The Active Antenna Systems will be required to handle more traffic than ever. It will be required to talk to more UE devices than ever before. How can they do this?

Like I said earlier, the antenna will have many more elements and the radio head connected to each element with a dedicated port that can handle the traffic.

If it's TDD, then the radio will talk and receive in the same port.

If it's FDD, then there will be an element for transmitting with a port to the radio head for that and a dedicated port for receive. All of this in the same antenna. This is what needs to be done.

So now you know that each element will have a dedicated port connected to it, that should be enough, right? WRONG! Now we need to control the element's beam.

The Massive MIMO Report

What makes massive MIMO special?

I'll tell you what, the fact that each sector will have over 32 antenna elements dedicated to talking to the UE device. It goes deeper, way deeper. The technology is awesome, let's look at the breakdown.

- For each element, which looks like an individual antenna, there is a radio interface, whether it transmits or receive. Think about it, the evolution from one antenna transmitting and receiving to 2, then 4, now 32 or 64 or 128 all concentrated in on direction! All on one sector, all dedicated to that group of users in that specific area. Talk about densification, 64 antennas all talking in one direction, working together to crank LTE through to all the users as a coordinated effort, like music flowing through a concert hall where the conductor trained the symphony to make beautiful music from 30 or more musicians all flow into one song like it's one big signal for everyone in the concert hall to hear it and enjoy it in their own way. You heard me, beautiful music, synchronized and streaming at amazing speeds. This is LTE flowing to the device in the same way, creating more bandwidth from the same spectrum. Awesome!
- Beamforming is another trick that massive MIMO will use. The elements can form the beams to talk to each specific user. That's right, the beam from each element will be shaped to "talk" to that user in the most efficient way possible. This is very complex, but the technology is here and being tested. What is beamforming? It's the ability for the antenna element to "shape" the beam to focus in one specific area at a specific angle to best talk to that UE device. It could be at 180 degrees, 90 degrees, 45 degrees, or anything in between. Why do this? To reduce interference to the neighboring beams and to dedicate that signal to that user increasing the bandwidth to that device. Isn't that amazing? I think so because now the BTS can dedicate that stream to that device and pass data at breakneck speeds while the neighboring element can talk to another user.
- Transmit diversity allows the UE device to grab the best signal and lock on to that signal. It can get the best signal as well as pick up the other signals around it as needed.

- Multiplexing is used for any MIMO, so many options. Multiplexing gets complicated in massive MIMO, but the new systems allow linear processing of all signals. This means that the performance of the system for each user goes up exponentially. While throughput may seem to be like that of the original system, it is the throughput to each user at the same time that makes the difference here. Beamforming and many radio heads can do this with the multiplexing gain achieved through optimal conditions.
- All of the above allow for multiple users to communicate with the same system in parallel. The sharing of airtime is minimal because each user or group of users will have their own beam to the antenna to pass data.
- One more thing, with the beam being direct it allows more power to get to the user. That means that the beam is concentrated vertically and horizontally like a point-to-point beam form a microwave dish to that particular user. However, the user still needs to talk back, but with aggregation, they could talk back through a local small cell. This means that the downlink will be huge, but the uplink is still average.

What is beamforming?

Antenna's that can control their own beam shape, what?!? Control the beam on demand? How can that be? Beamforming is a little more complicated than that.

First, a quick, high level, history lesson.

I don't know how familiar you are with antennas, but they must be installed correctly. You could physically tilt the antenna a few degrees to match your coverage. It's like azimuth, that must be aligned properly for coverage. Older antennas were installed with a set "up tilt" or "down tilt." They were fixed in tilt and azimuth. So, what they saw is what they heard, based on the antennas fixed pattern. The antenna pattern would determine what the antenna could hear and talk to. That was it, very simple. I know, there is gain, but for the sake of argument, let's say they would talk to UE devices in their specific coverage area.

Then there was evolution! There were new ways to control tilt. CommScope had RET, Remote Electrical Tilt, for this purpose. I think it was a good idea, but it's still a physical system. Basically, if I understand this, it's an actuator that can change adjust the tilt + or – 3 or more degrees. However, it opened options to the end user, the carriers, where they did not need a tower crew to adjust the tilt. Pretty cool!

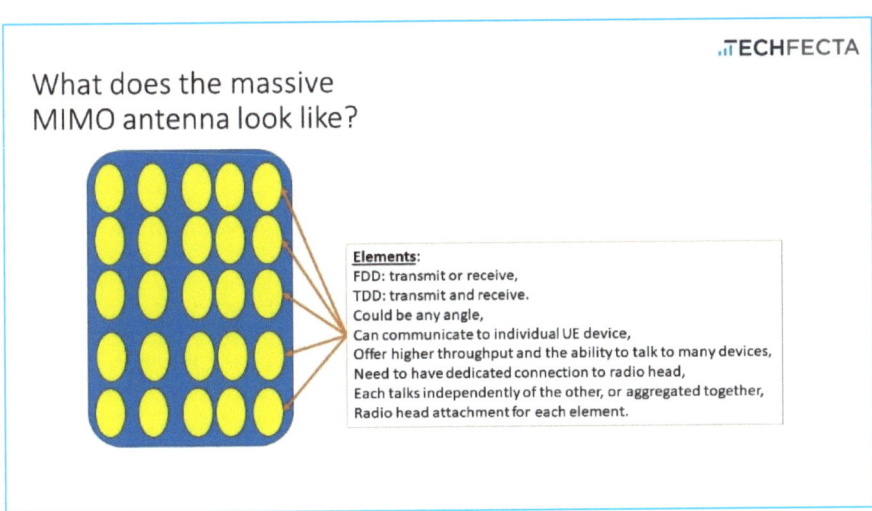

Figure 4 Massive MIMO Antenna

How does it work?

Now a new type of revolution, beamforming!

With beamforming, none of the physical alignment goes away; we still need the proper tilt and azimuth to get started. Beamforming is done by very smart antennas, but the carriers did not have the corner on this technology. As a matter of fact, the Wi-Fi vendors have made significant advances in this technology. They did a great job getting 802.11 to do this. The BTS controls the beam from there so that the antenna can do its thing. Again, they all must work together to make this happen.

Where did this idea come from? Don't let the carriers or OEMs fool you; it came from Wi-Fi. In fact, I believe one of the pioneers in beamforming was Ruckus! That's right, the carrier-grade Wi-Fi OEM. Also, I must give credit to Linksys for putting the technology in their home Wi-Fi routers. Awesome! Thank you to Network World for making a video on this, (link is below in "Learn More" section).

Massive MIMO puts that on steroids. It takes the signal, both ways, and focuses on a user. If you have 64 by 64, then, in theory, you can focus on 64 individual users on that antenna. The idea is to hear users you want to hear at any moment. This allows the radio to talk to specific users simultaneously without sharing precious spectrum. How can they focus? Beamforming is how they do it. They employ a technique called 3D beamforming that dimensions the signal from that element in 3D, 3 dimensions. Beamforming will focus the beam on the specific user.

Now the carriers are asking the OEMs to take it to the next level. It is the cornerstone of making massive MIMO even more useful.

You see, massive MIMO relies on the beamforming technology to make it more efficient and push even more bandwidth through it! It is a crucial factor, like carrier aggregation. It all must work together.

Now, by controlling the beam to match the user's antenna, it becomes more efficient in several ways. Signal strength helps, but now the spectrum stream can be dedicated to that specific user the duration of the conversation. Not only the best signal possible but a dedicated conversation with that unit for a limited time.

Massive MIMO takes this to a new level. Now the angle of delivery can be controlled. WOW! That is specific to the user based on each element. Assuming it really works that way, the antenna will have to be smart. This is called 3D beamforming, looking at all 3 dimensions. It's steering the beam to match the end user, basically taking a lobe that focuses all 3 dimensions to the user's antenna. Almost like a microwave shot to the antenna. Beamforming is shaping the beam to match that of the UE device. The antenna will narrow the beam so that it is only talking to the device or devices that you want it to talk to and not the surrounding units.

That makes the antenna elements very efficient. This increases the number of devices the antenna can talk to as well as increasing the throughput to each device with a dedicated stream from each element.

Remember that the antenna will need power now since the radio heads are in the antenna and the elements need to be agile. The elements will control the beamforming, and they need to have control signals sent to them. It's a whole new level of technology.

MIMO helps to utilize beamforming by using the radiators in the antenna to focus on specific users and not "hear" everything. That's the key, listen to what matters and forget the rest.

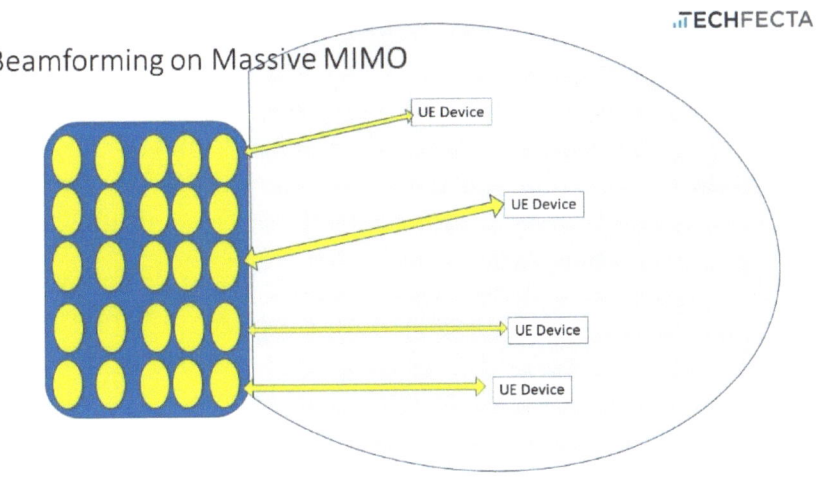

Figure 5 Beamforming 1

What spectrum does beamforming work in?

As far as I can tell it works in any spectrum. There are arguments to put it on the higher spectrum. I mean it works in Wi-Fi. It is being tested in 600MHz with T-Mobile, it seems to be working there. There doesn't appear to be any limitation based on spectrum. The sub 6GHz spectrum may have too much power, so it may need to be limited. The higher bands, over 6GHz, will benefit more from beamforming.

I think the carriers will get it working on all their spectrum; they need to get it rolling. Whether it's FDD or TDD, it will be the foundation for massive bandwidth to the end user. It's a matter of how to reach the user.

Who will use it? (Looking at the USA only)

You mean after all the Wi-Fi vendors? They are already using beamforming and MIMO because it really helped throughput. Then the carriers are all going to use this. It means changes at the sites. New antennas with upgraded BTS systems, and even backhaul and fronthaul upgrades. This all must be upgraded.

They all want it though, they all want to serve the public. The question is how?

For instance, if you read Verizon and AT&T press releases, you see that they intend to deliver high-speed broadband to homes vial cmwave and mmwave. They intend to open new markets to the end user that would, in my opinion, compete directly with the cable company's model. Get ready Comcast; you will have competition! They will have controlled rollouts for the fixed wireless market. They will rely on massive MIMO for that solution first. I get the feeling they both will follow the roadmap to massive MIMO for the mobile systems. From my perspective, they don't seem to be in any hurry to upgrade their existing sites to massive MIMO. Maybe the competition can help them change their mind.

Sprint is making plans to deploy the massive MIMO in 2.5GHz in 2018. This will be a huge stepping stone to 5G for them. Once massive MIMO is up and tested, they should see dramatic results in throughput and performance that other carriers can't do. That's going to be a gamechanger for Sprint. They could finally have a technology edge on the competition.

T-Mobile has more spectrum now that they won the 600MHz auction, and I think when it's ready to be deployed. However, how will they do it in the

600MHz spectrum? I don't know, but John Legere has an answer. I certainly don't want to speak for him, but I know once they get this technology and have faith in it, they will go crazy to get it out there. T-Mobile is already making plans to deploy massive MIMO in 600MHz. They know that they have to get one up on the others. This is a way for them to get national coverage and improved throughput that will take them closer to 5G.

The cable companies, meaning Comcast, should be eager to do this, but I don't see them aggressively doing anything with it. I see them investing more in Wi-Fi. They see the writing on the wall; they see LTE passing Wi-Fi speeds, yet, they don't seem worried. They even saw John Legere mention how he was going to go after them, yet, they seem very relaxed. I don't get it, but they are very successful, and only a few cable companies are debt-ridden, so they seem to be doing something right. They have a corner on the suburban markets for sure, but the urban markets may start falling behind.

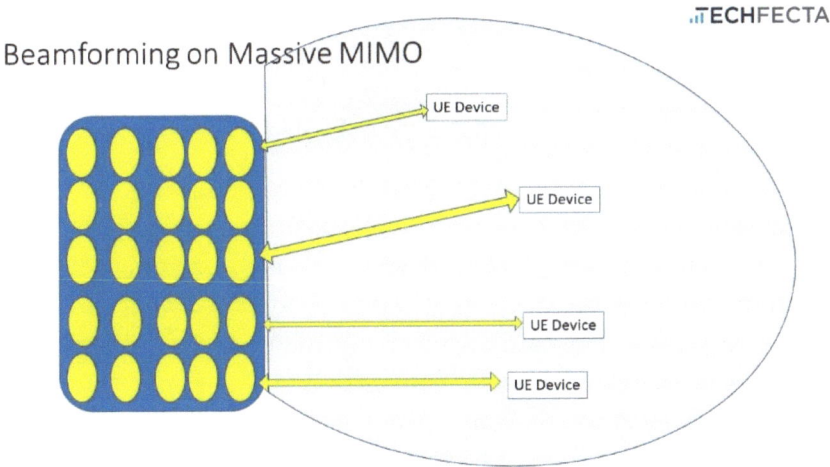

Figure 6 Beamforming 2

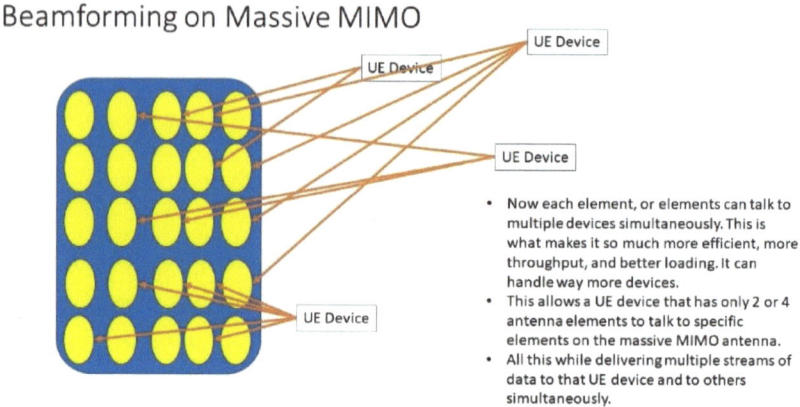

Figure 7 Beamforming 3

Why cable companies should pay attention.

I would heed what John Legere says about cable companies, even though Comcast's approval rating is up. T-Mobile already proved they could change the stubborn wireless industry. He singlehandedly destroyed contracts and lowered costs and built a following for unlimited data plans. I believe that he could do the same for cable subscribers, mainly because millennials rely more on their devices than ever.

Personal story, my son was living in Ann Arbor Michigan, in a program at the University of Michigan. When he was there, he didn't have a cable subscription. He did have an internet connection to them and Wi-Fi, but he complained about it all the time. As a young single man, he relied on his laptop for all his video viewing, movies, and YouTube. He didn't watch TV on cable; he watches it on his TV with his iMac feeding the TV. He didn't rely on any cable box or anything, just Netflix and YouTube. My point here is that millennials look at broadcast completely different than we do. They know that all you need is the internet, then you can watch whatever you want. He didn't care how it got to his apartment if he had Wi-Fi inside and it was fast. He would have been just as happy getting it from his iPhone instead of Comcast, but AT&T was too slow and didn't have the best coverage in that area. So, he got Comcast, and it worked fine.

The moral of that story is that the new generations could care less how they get internet access. All they want is a connection. They rely on apps to do the rest. Whether it's entertainment or voice, it's an app. They use Skype like we used to use a phone. It's an app that matters.

What about you? What do you rely on? I know one thing, in the next 5 years you will rely on beamforming and not even know it. Like massive MIMO, and carrier aggregation, you will use it all, but not be aware of any of it.

Learn more:

- https://www.networkworld.com/video/64563/what-is-wi-fi-beamforming-and-mu-mimo
- https://www.smallnetbuilder.com/wireless/wireless-features/32329-does-beamforming-really-work
- https://kb.netgear.com/31299/What-is-explicit-beamforming-and-how-does-it-work
- https://theruckusroom.ruckuswireless.com/wi-fi/2015/05/01/making-the-most-of-multi-user-mimo/
- https://www.ruckuswireless.com/rucktionary
- http://blog.3g4g.co.uk/2012/10/3d-beamforming-and-3d-mimo.html
- https://www.slideshare.net/khalidhussain359778/3d-beamforming
- http://www.rle.mit.edu/ncrc/wp-content/uploads/2015/05/Phil_5G-Day-at-MIT-5-8-2015_Nokia.pdf
- http://www.philly.com/philly/business/technology/t-mobile-john-legere-announces-tv-alternative-comcast-streaming-internet-time-warner-dish-20171213.html
- http://www.telecomhall.com/what-is-antenna-electrical-and-mechanical-tilt-and-how-to-use-it.aspx

The network matters!

With 5G you hear a lot about SDN and VNF, but this is all going to happen with OTS equipment. When looking at the 5G specifications, they talk about the network being part of 5G. That means new functionality and automation will be needed for the network. All of this is well and good, but to get the broadband that massive MIMO systems will need, the backhaul has to be upgraded.

Just imagine that the connection has to increase 10 times or even 100 times what it is now. That is the expectation. Many macro sites may have a 1Gbps connections. Some have 10Gbps backhaul. With the growth of massive MIMO, we expect to see 100Gbps as the new normal.

We'll see the first Terabit links to cell sites, probably in 2020 when 5G starts to take off. That is more than one strand of fiber! The connections will continue to grow, especially in urban areas. If this happens the right way, the need for Wi-Fi may start to decline as the efficiency of 5G systems starts to surpass Wi-Fi. I know it won't go away, but it will be less of a factor for throughput.

What about the backhaul?

It needs to be upgraded to handle more data since we are now going to increase the broadband throughput. 1Gbps backhaul may not cut it because it may need to be much more. What does this mean?

- Better routers at the sites that can handle the higher throughput and lower latency.
- More strands of fiber for each site. It may take more to handle the uptick of traffic.
- Aggregation of the fiber carriers so that it looks like a massive pipe of over 10Gbps.
- Improved network slicing functions for the router to perform more functions than before.

The need for bandwidth only increases with massive MIMO, exponentially. The bandwidth will increase so that all the users have a much higher throughput. It's going to take more than 1 pair of fiber to handle the traffic.

This has major implications on the performance of the site. If you add multiplexors to use the existing fibers, you will increase latency and increase

CapEx. If you add fibers to increase the bandwidth, you increase OpEx. This is where the balance comes in where the payback must be better than the investment in the backhaul and new equipment.

What about the fronthaul?

The fronthaul is the fiber, (or wireless), the line between the BBU and the radio heads which now should be in the active antenna.

This will require more strands per sector so that the data can be sent to each sector. Look at it this way, instead of dealing with one antenna with 4 to 8 elements; now you are dealing with one active antenna that could be simultaneously sending data to more than 64 elements in one concentrated antenna system. That is a lot of data! In theory, it could be more than 1Gbps per sector! WOW!

Let's not forget the latency. When at a site the latency is no big deal, but if we try to configure a CRAN system then the latency may be an issue. There is a lot going on between the BBU and the massive MIMO radio head. Remember that the BBU is processing traffic and overhead for 1 radio head, but it sees 32 or more ports in the radio head, making it look like 32 or more radio heads. That is a lot to process.

At the site, be it tower or rooftop, there will be a critical engineering facto that needs to be thought of, the fiber and power run between the BBU and the radio head. It's more than what they have been using. The radio head needs power, possibly more than the original radio head, and will use a lot more bandwidth. It has to be improved.

Fronthaul will need to be:

- Improved for more bandwidth and data.
- Lower latency than before.
- Perhaps more strands of fiber.
- Perhaps more power needed at the radio head than before.

How will the network meet the demands of 5G?

It will have to be a combination of equipment and connections. The connection will need to be more than the fiber that is deployed. Wireless backhaul connections need to improve drastically. They need to allow for growth, and they need spectrum to do that.

The equipment needs to become faster and smarter. Network virtualization will need to improve. It is already making great leaps.

The thing that may hold back most of this is that the equipment is expected to be cheaper than ever before. If equipment is cheap, there's not a lot of incentive to improve.

The networks need to improve, meaning that routing will have to get more creative. It's going to take more than one backhaul to make this happen efficiently. We need to have more redundant links, but not just for redundancy, but for sharing of bandwidth. Routing will be virtualized assisted by artificial intelligence helping it along the way. AI will make the routing more efficient than ever. It should lower bottlenecks and make the networks as efficient and fast as ever.

What about the extensions of the macro sites?

It's going to be hard to look at just the macro sites. With CRAN and small cells playing a big part of the wireless ecosystem, we need to break it down.

Will CRAN or C-RAN be a massive MIMO system?

The jury is out because of the high fronthaul bandwidth needed to make the radio heads do massive MIMO. The challenge is to get massive MIMO to work on CRAN through the fronthaul using the fiber they have at the site now. They may need some type of multiplexor, or it will have to be a cloud RAN where the cloud runs some of the BBU functions.

The fronthaul needs enormous bandwidth because of the BBU and radio head communication has a lot of overhead as well as the traffic data. Now that the BBU is talking to 32 or more elements at any given time through the radio head it has a lot of communication, control, and commands to send out. Imagine the BBU needs to simultaneously talk to 32 smaller radio heads that are all cased in one unit. Like I hinted at, this would require massive bandwidth for the overhead. Quite a challenge because at the macro site they have many fiber pairs running directly from the BBU to the radio head full of power and fiber pairs so that they can communicate properly. CRAN has to figure out how to do all of that remote.

The OEMs will figure this out, probably using some type of multiplexor to get more data through existing fiber. I also think that the active antennas will start to have more computing power in them. There may be a controller that is an extension of the BBU put at the CRAN site. OEMs have been trying to change the latency requirements between the BBU and the radio heads to make the CRAN work better. Massive MIMO makes that very complicated. But hey, they are full of smart people who will figure it out, someday.

It is essential that we know that a Centralized RAN and a Cloud RAN are essential parts of the 5G system, so hopefully, they can do massive MIMO. As you all know, massive MIMO is a huge stepping stone for 5G. I would think we need to get massive MIMO to work at a CRAN site. How can we do this?

Maybe the MEC solution will help, or a cloud service. Something where the BBU can hand off some of the processing to the local radio head site. Normally we look at the MEC, Mobile Edge Computing, the solution to help create low latency. It may be the solution to offload some of the BBU functions to the local site. It makes sense.

Low latency will still be an issue, so the BBU location may have to be closer to the radio head site than previously done with the normal CRAN system without massive MIMO. Why? Even though it is light running through the fiber, it takes time to travel across town, the state, or anywhere. Hence, that is how MEC may be able to save the day! Direct routing would be a key factor instead of running everything back to the core. The routing needs to be as short as possible so that latency is as low as possible so that the commends between the BBU and the radio head are delivered in a timely manner.

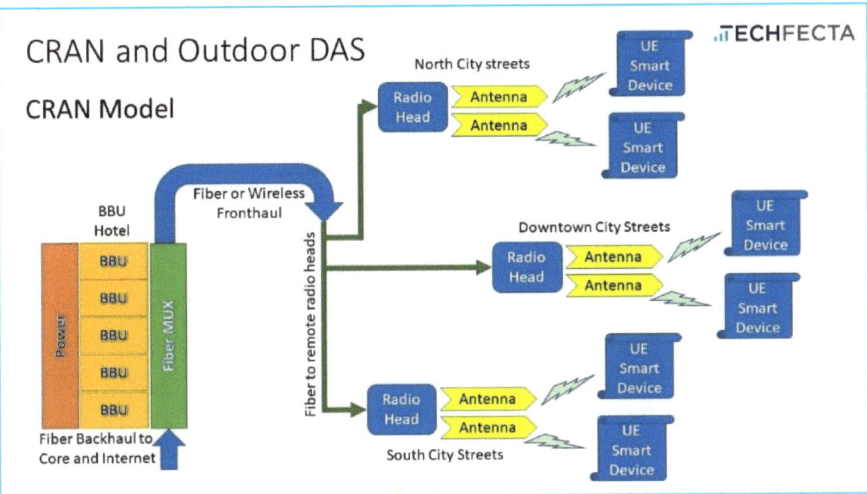

Figure 8 CRAN and oDAS

Resources:
- https://wade4wireless.com/2015/07/27/small-cell-fronthaul-and-odas/
- https://wade4wireless.com/2018/04/15/would-you-deploy-small-cells-or-cran/
- https://wade4wireless.com/2016/02/08/densification-breakdown/

Will Small Cells have Massive MIMO

The small cell is not well equipped for massive MIMO. They already do some types of MIMO, the may have 2 or 4 antennas that allow it to use MIMO to improve throughput and receive signal levels, but the reality is they won't have the power or size for MIMO. Let's not forget the money. The idea of a small cell is to alleviate what the macro sight can't do.

What can't the macro site do? It can't fill certain areas. Small cells make it cost effective for a carrier to put a site in the basement of a bus stop so that there is a better signal there. Let's not forget the offloading. That is another thing that the small cells do quite well. They offload data, just like Wi-Fi and someday CBRS cells will do.

The problem right now is that small cells need to be cost effective for the purpose they serve. They need to provide a small solution. If you know much about the outdoor small cell, it's that they were not deployed to the masses as we all thought. The reason was that many of the site acquisition and backhaul costs were nearly what a macro site is. That defeats the purpose of deploying quickly and easily.

Indoors they still serve a great purpose to do all that they were expected to do. The great thing is that there are more backhaul options inside. If a company wants a small cell, they could buy it themselves and run it through an existing backhaul. Why make it complicated.

I don't see small cells using massive MIMO, but they will continue to use MIMO as long as it's cost-effective. Think about 5G, if it's as efficient with the spectrum as we all hope it is, then small cells will go way up in value. If the small cell can increase data throughput to the point of a small macro site, then I would say it has potential.

What about DAS systems?

Looking at DAS systems today, they have been using more and more digital fronthaul to talk to the radio heads. It's been the trend. We can't look at traditional DAS as a way to move forward. DAS systems today for LTE and 5G are all digital. Compare them to the CRAN systems that I overview above.

However, DAS services a vital function in the ecosystem when looking at large venues like stadiums. While the carriers are hoping that macro sites can supply the necessary coverage, it can't supply NFL stadium loading. That is why they put DAS systems in stadiums. It has to cover 50,000 users simultaneously. This is a key element for massive MIMO. I can see massive MIMO active antennas starting to be deployed at stadiums to improve loading and densification. It could make it more efficient than ever.

When massive MIMO systems are deployed in stadiums, they will improve the loading to the users during that big game. It should allow the carriers to do more with fewer antennas. Remember that the antenna is active, so there will be no radio heads or coax. It should allow for a cleaner installation farther from the users. Let's face it if done properly. The user's experience should improve by leaps and bounds.

The Massive MIMO Report

Will Massive MIMO be in the UE device?

The UE probably won't have more than four antennas in it, if that. They may have 2 to 4 receive antennas, and two transmit. The thing is though, with massive MIMO and the way it should work, an element or 2 can focus on one US device freeing up the other to talk individually to other devices. This will improve throughput and lowers latency.

The bottom line is that it will help the UE in 2 ways. The UE will get more bandwidth, especially as they add more antennas internally, they already have two receive in most devices which allow for better downloads. The massive MIMO will be able to talk directly to an individual device and lower latency. This will help the response time of the device, making the device seem more responsive and once again, quicker! Now it is up to the device makers to speed up the internal processing speed and improve the memory in each device so that we can enjoy the new low latency services.

This may change as devices get better, smarter, and larger. The thing is that smartphones are cranking a lot of data now, how much more will they need? I don't think the bandwidth of the device is as important today as it once was. They seem to be doing fine. I think the carrier needs to eliminate the bottlenecks. What this means is if you are around 1,000 other people trying to look at the cat video, like you, the cell site needs to satisfy all of them at the same time. This is where massive MIMO can work its magic. You don't have to wait for Biffy and Muffy to download the video first, you can all do it simultaneously. This is the magic that people will appreciate, even though they will not see it that way. All they know is that they get the bandwidth they so desperately wanted when they needed it. Apparently, they need it when a new kitty video is released.

Another thing I believe we will see is the resurgence of the laptops and tablets with carrier ships in them. As carrier systems, improve and unlimited data is commonplace. Why would they start putting business machines on these systems? If you would go back to the day when you could use Wi-Fi or a carrier, especially if its 5G, why wouldn't you do that? If you're paying for a carrier's network anyway, make the most of it. You could continue to tether the device as I do, but wouldn't it be nice if you would have the chip on the laptop already? Why isn't Qualcomm working on this today?

What changes will tower companies see at the site?

This is going to require new equipment at the site, no matter how you look at it. The equipment that is there now will need to be upgraded or replaced. Most likely, some of it will need to be replaced. It would make sense to have 64 or more radio heads so that you would replace the existing antennas and radio heads with an active antenna. You would remove all coax and run fiber to the antenna. You would need a BBU that can kick ass with processing power and bandwidth. You need to improve your backhaul, which means a better router CRAN quality equipment at the site. You need to improve your fronthaul between the BBU and the active antenna.

To do this 5 years ago we would have needed more equipment on the tower. It would have been heavier and ugly. Now, the equipment is getting smarter and smaller, which is great because it's where the industry had to go.

Smaller form factors are key because the towers are loaded, and the rooftops need to conserve space. The real kicker for the carriers is the fact they don't want to pay any more rent than they have too. It always costs money to make changes at the site, that's how site acquisition companies stay in business. Any change to a site generally requires new drawings, lease modifications, and so on. The good news for carriers is that it will not raise the monthly cost of the lease if the equipment is smaller and lighter. It also saves them from making structural changes to a tower, which is very expensive. Most companies want to build new towers, but there are restrictions that may not allow doing that in most places. So many towers must be improved structurally if they are adding more equipment. Therefore, the carriers are doing all that they can to have smaller and smarter equipment.

If the OEM can make equipment smaller and smarter, then the carrier won't have to worry about additional OpEx that could sink their business case.

All the same, the site acquisition companies will be busy during the massive MIMO deployments. Many leases will need to be modified. The changes on the tower will be enough to make the tower companies figure out how to charge more for the equipment on the tower. I list a major reason below.

On the Tower:

Just to lay it out for you what will be on the tower:

- Active antenna
- Fiber to the antenna
- Power to the antenna

What will be different from most LTE deployments:

- No more Radio Heads
- No more coax cable

This means that all the weight will be at the end of the mount. That means that the mounts will need to be strong enough to hold the active antenna, which will include the radio heads and the radiating elements.

The major determining factor will be the spectrum that the antenna works in. Let's look at T-Mobile and Sprint.

T-Mobile is at 600MHz for their new spectrum. They intend to roll out 5G and massive MIMO in that spectrum. That means that the lower frequency needs bigger radiating elements and since it's FDD, it needs both transmit and receive radiating elements. They probably can't be shared. So, the tower will have this large and heavy antenna on it. So, it won't have radio heads or coax, but it will have a lot of weight. Since this is a new deployment, it's going to add a lot of weight to the tower and the mount. Major thought has to go into the structure of the tower and mount.

Now, let's look at Sprint. They have 2.5GHz and TDD. That means the higher spectrum allows smaller radiating elements, good news for structural on a tower. They also have TDD which means that they can transmit and receive on the same smaller elements. This is all good news for Sprint. While the structural will need to be looked at, it will be an easier deployment than what T-Mobile had ahead. While structural still need to be done on the tower and mount, it will be more forgiving than a massive FDD 600MHz active antenna.

One more thing to think about, at this time there won't be many multiple spectrum antennas. As far as I know, they haven't been made yet. I have no idea whether the OEMs are working on combining multiple spectrum ranges into a single antenna. It will be an engineering feat to get all the spectrum the carriers have into one antenna. Lucky for Sprint they have over 100MHz of spectrum crammed into 2.5GHz.

On the ground:

The ground should not see much of a change. They have the same cabinets and will try to keep it all in the same size BBU. They may need more power, but other than that the ground should remain relatively the same for massive MIMO and 5G.

Let's not forget the backhaul. The backhaul will need to be upgraded past 10Gbps, maybe to 100Gbps. This won't affect the ground footprint, but it will mean that the routers have to be upgraded. Chances are the BBUs may need replaces, and there may be an upgrade in power supplies. However, if looking at the physical footprint, no change.

If fiber has to be added, how do they get it to the site? Chances are it will be buried, and they may have to dig up the original to add to it. It all depends on how much growth the original fiber provider planned for.

One more thing, for the sites that are fed with microwave, that will not be sufficient. Either the microwave has to be upgraded, or they will need to run fiber to the site. That is the problem with older microwave radios; they were very limited. Chances are, if they need a new radio, it will need new spectrum and a new design. The old spectrum probably can't do over 10Gbps unless it is mmwave. The good news is that the new radios are generally smaller. The bad news is that range is much shorter. So, what do you do? The chances are good that the carriers will run more and more fiber.

Will Utility costs change?

Well, this is tricky. The original intent was to use less power. I don't see that. This new technology will use more power, but you get a bigger bang for your buck. So, it may take more power at the beginning because the processing power and the number of radiating elements go way up. It's crazy!

The other utility is the backhaul, the fiber costs if you call that a utility. It will go up because the demand is higher and necessary. The carriers may try to multiplex the signals, but this may not work, it may take too much money up front when the new fiber needed may be easier to implement. Remember that massive MIMO should allow 30% to 60% more data through to the cell site. Maybe more. It's going to need a large backhaul to accommodate the increase in broadband.

Massive MIMO Tower Work Overview

Overview

Massive MIMO is the next wave of deployments that will lead to 5G at the macro site. You can put massive MIMO anywhere, but it makes the most sense to put it at the macro site because of the cost. It is not cheap to deploy, but once it is in it add so much value. Why put it in? Because the densification increases exponentially. How does it do that? It has extreme beamforming techniques that allow the antenna to focus on 1 or 2 users. Why is that important? Because the power and data can be focused where it is needed the most. Now, imagine it can do that to over 1,000 users simultaneously. It would be a multi-user MIMO, MU-MIMO. This is going to exponentially improve coverage and the number of users on one sector at any macro site. That is why it will be a crucial differentiator in 5G. It will do amazing things for LTE, so 5G will increase the value even more. That is what massive MIMO will do for the carriers, and that is why they want it for their systems.

It's important to understand that this is a big step towards 5G for the macro site. It's something the carriers will do everywhere it makes economic sense. They will start in the urban areas then work in the rural areas. They may not put in massive MIMO at all macro sites because it's an expensive upgrade. It is a lot of costs to put on a site where there are few customers.

What is Massive MIMO, really?

When companies refer to massive MIMO, they mainly are talking about the antenna. It's more than that really. It is the way the antenna and the radio heads transmit and receive. It is a smarter antenna and a lot more radio heads. Imagine now that an antenna has 32 radiators in it. Now, imagine it has 32 by 32 radiators in it. Imagine that they are all small and that they all can control their beamwidth in all directions.

Think about what I said above. It would be too much to have that many radio heads outside of the antenna. The radiating elements in the antenna need to be smaller since there are so many. However, if you could use smaller radiating elements and tiny radio heads and cram all of that in one antenna, it would become an active antenna. That is what most massive MIMO systems will be, an active antenna on the tower.

So the antenna is active, which means it has electronics in it. That means that the antenna will need power. So feeding the antenna will be a power cable and fiber jumpers. Coax is not needed for an active antenna, just fiber, and power. What's the point?

This changes the design from the BBU up the tower to the antenna. You're still running fiber and copper up the tower like most LTE systems today, but you need more of each.

The fiber needs to feed more than 1 to 4 radio heads. It has to feed more than 16 or even 256 in one antenna, small radio heads, but radio heads all the same. So you could have at least 3 fiber pairs to the antenna and possibly up to 16 fiber pairs.

To power something like that, chances are you need more power, so each antenna will probably need a dedicated power source. These things need power to run, more than a few radio heads. So that is a given.

Back on the ground, the BBU must be upgraded or replaced so that it can control the active antenna. It also has to process more data than ever on the path to 5G. It also has to send the overhead to control the power, beam width, radiating elements in the antenna. That is a lot of overhead. This means that the cables need to be upgraded or replaced between the BBU and the antenna.

The entire system will need to be upgraded. This means a lot of work at the tower site. This is good for the tower worker. Lots of work through 2020.

What about the tower work?

If you're reading this section, then you care about the tower work. This is a mystery to most right now, but let's go over some commonly asked questions. Basically, on the tower, the new equipment will replace the old equipment. The radio heads and the coax connecting them, and the antennas will need to be removed. They will all come off the tower.

This means that you need to remove the systems you put in about 5 to 10 years ago. The radio heads, coax, and antenna. You probably want to take down the assembly if you can. However, chances are the carrier will only upgrade one band, not multiple bands. This means that you may have to remove a partial system. Massive MIMO systems that I have seen only radiate on one band. The system planning will need to take this into consideration.

What if you swap?

If doing a swap, take this into consideration. The system on the tower probably needs to be replaced altogether. This means that the antenna, the radio head, and all the coax will be removed, and a new active antenna installed where all that equipment was. This is a different system, but a simpler one on the tower. The weight change may be minimal, but all the weight is the antenna. The antenna has all the electronics in it. The weight allocation will no longer be evenly distributed between the radio heads on the back and the antenna on the front. Now, all the weight will be at the antenna. All the weight is going to be at the end of the mast on the mast holding the antenna.

In the old system, the weight was distributed across the mount, from the radio heads in the back to the antenna on the end of the mount. This distribution of weight across the radio head of the radio heads, coax, and the antenna was spread across mount from the end back into the tower. Today's equipment will be smaller, but we're putting more and more into one piece of equipment. The new equipment may weigh the same, maybe more, but it should be close. Of course, this depends on the spectrum and design. More on that below.

If this is a swap, then consider the mount. With TIA-222 Rev H coming out, the mount now needs to be analyzed in most states. This means it may fail. Hell, it could have failed before you even did any work. Some carriers didn't bother to do a mount analysis when adding the additional equipment. Now they are forced too. You may have to do it even if this is a rooftop. So, you may have to replace the mount. That is a lot of work. Plan accordingly.

If the mount is good, then chances are you will have to replace the antenna mast. The old mast was good enough to hold a 50lb to 100lb antenna, not the antenna may weigh 150lbs or more. All the weight will be on the antenna mast whereas before you had the weight distributed between the antenna and radio heads. The radio heads were mounted farther back on the mount. This meant you had a better weight distribution. Now it is all on edge, on the mast holding the antenna. This changes things considerably.

What is it's new?

If this is a new install, then you may be putting in new mounts altogether. It won't matter about like the swap. The new mounts should have been picked out in site acquisition. The tower has to be made ready for the mounts. It will be a major installation, but nothing you can handle as long as you planned ahead.

If you're going on an existing mount, again, the site acquisition team should have already known if the mount will hold the new antenna or not.

While a new install is a lot of money for the carrier, it will be very clean. New installations can be challenging, but there should not be any downtime. It is a new clean install, in theory anyway. There are always challenges that the installation teams will run into. No tower work is without challenges. The tower teams have to assess the situation, adapt, and overcome the problems while on site if possible. That is the nature of the business.

Who decides what mount is safe for massive MIMO antennas?

This should all be determined before the climber gets to the tower. Here is what should happen.

First, site acquisition will do 2 things. They will do a structural on the tower and the mount. Both need to be done according to TIA-222 Rev H.

If the tower needs to be modified, the should catch it and put it in their report. This is going to be a considerable charge if the tower needs to be upgraded. Chances are it's a different crew and skill set doing structural modifications.

If the mount needs to be upgraded or replaced, then site acquisition should find it and report it. The mount analysis report, which should include a mapping of the mount, should explain what needs to be done.

Second, the structural engineer should make recommendations for making the modifications to the tower and/or mount. This is where the design should be done. They should know what needs to be added to the tower or mount to make it safe. This is all going to be caught in the structural and mount analysis. The structural engineer will use that analysis and the information on the tower and mount so that the proper changes can be made.

What about the cables?

On the existing systems where the radio heads are on the tower, there cabling is similar, minus the coax. The coax will go away with an active antenna. The fiber and power will connect directly to the antenna.

However, most systems are going to have to replace the cable that is there. Why? Because the fiber is not sufficient to run most new systems. The amount of data passing through the new systems is going to be 10 to 100 times as much as it used to be. The fiber requirements will increase because the amount of data passing from the BBU up to the antenna has grown exponentially.

One more thing, chances are good that each antenna will need more power than before. They will each require power from the ground. They will require more DC power to make the MIMO massive. It takes power to have all of this functionality and power all those radio heads in the antenna simultaneously.

So, there you have it, the hybrid cables will have more fiber and more power. This cable could is 3 to 10 times bigger and heavier than the original hybrid cable. This is added weight and complexity to installations.

For the tower, the hybrid cable still needs to be shielded. The connectors should be on the cable already, but that only makes the installation more of a challenge. In the old days, the tower climbers had to have fiber termination skills. While the new cables will be pre-connectorized, chances are the rough treatment at the tower will break off fiber pairs and connectors. I know what goes on when trying to pull the cable up the tower and secure it into the cable tray, it takes brute force to do that.

Not only a tower but can you imagine trying to install one of these up a monopole center where there are already over a dozen cables. How do you deal with that? It's going to be a challenge over the next 5 years. I would imagine the tower crew will need to have a fiber termination kit just in case connectors or pairs get destroyed.

Is it bigger or smaller? Size and weight matter!

The active antenna will be heavier than the older antenna, but from what I have seen it is not much taller or wider, but it is deeper. This varies from OEM to OEM and of course the band it operates in matters.

For T-Mobile, who is rolling out their new 600MHz system, the antennas are going to be massive. Not just because it's massive MIMO, but because at 600MHz the antenna is larger than Sprint will have at 2.5GHz. The lower the frequency, the larger the radiating elements.

Also, for FDD systems, they will probably need a set of transmitting elements and a set of receive elements. This plays a vital role in the size of the antenna. The 600MHz FDD system will need 2 sets of elements, one to radiate and one to receive. It adds complexity and size to the antenna. Whereas the 2.5GHz antenna

With Sprint rolling out massive MIMO as a predecessor to 5G, they will be able to this quickly and efficiently with 2.5GHz TDD because the spectrum and TDD allow the antennas to be a smaller form factor. It won't be easy, but the antenna itself will not be a problem compares to 600MHz FDD which will be much larger due to the lower spectrum and FDD design.

Size and spectrum go hand in hand. When doing this work or bidding on it, you should know who the customer is. Not the OEM or GC you're working for, but the end customer, like T-Mobile or Sprint. The fact that they have spectrum at opposite ends of the spectrum will determine how you mount the equipment. It will also give you an indication of how they do things.

How Will TIA-222 Rev H affect Massive MIMO Tower Work?

What is TIA-222?

TIA-222 is the standard used in the industry to set the standards for towers and communication structures in the USA. TIA and ANSI develop standards for structures. TIA-222 is for Structural Standard for Antenna Supporting Structures and Antennas per the TIA.

- TIA Telecommunications Industry Association
- ANSI – American national Standards Institute.

Why does Rev H matter?

It might make sense to let people know how TIA-222 Rev H. For 5G it may or may not matter, except for the fact that new antennas and radio heads are needed. However, for massive MIMO it makes a huge difference.

Why?

Because the massive MIMO antenna has the radio heads in the antenna. Yes, this changes the dynamic of mounting the equipment. Specifically, the mount that holds the antenna. While we all think we need a new mast to hold the new antenna, it is much more than that.

Rev H has added a section that addresses the mounts, and that is what this article is about.

Why now?

First, a short history lesson. Many carriers thought it would be a good way to save money by installing the smallest mount and masts to hold the antennas as possible. It makes sense, right? Why pay more than you need? I'll tell you why growth and expansion. I remember when carriers would put in cheap monopoles to hold the equipment they needed. I can also tell you about how many bent over, collapsed due to loading. That's why TIA became so important. It set standards, and the states required that the standards be followed.

Many carriers would save $50 to $300 per mount and mast per tower. If you're deploying 30,000 towers, that adds up to a lot of money. However, did you ever hear the statement, "pay now or pay later"? You eventually must pay. Sometimes it pays to do the work up front to avoid downtime and excess costs later.

How will this affect new deployments?

Now, back to rev H. You may think that Rev H is more about the tower, but it's not, it's the mount that is looked at here. Section 16 is the new section dedicated to loading the mount, the design, the requirements. Before the mount was not looked at by everyone. Now, with rev H we must look at the mount, the design, and do a structural on the mount. The mount is key to holding the antenna. Massive MIMO will have heavier antennas. This could be a game changer for the carriers as they expand.

TIA-222 Rev H was implemented in January of 2018, and it's going to play a big part in rollouts. I would bet that several carriers will need to replace mounts. It could take the site off the air of the mount, and the platform must be upgraded. It is also going to give structural engineers a lot of work.

This is going to be a game changer in deployments as all the carriers will need to do a complete mount analysis for their towers. Documentation will be required.

Can you imagine the work it's going to take to change a mount during an upgrade? It will take more time to do this. Would it make sense to just deploy the new equipment on a new section of the tower? No, rent increases would be an issue, and the tower may not handle the loading of another platform. So, the upgrade is the way to go. They will do an analysis of the mount, map it, then upgrade as needed.

There is way more to Rev H, but the mounts will be the focus of more carriers site acquisition. The mount is where the antenna is being changed or added. The antenna models are changing, and the radio heads will be incorporated into the antenna, so things are changing in the preparation of 5G deployment.

How does this impact 5G?

Because 5G will need new antennas. Specifically, if the carriers go to massive MIMO. Like I said earlier massive MIMO would have the radio heads in the antenna, so all the weight will be in the antenna. This changes the loading to the antenna. While it reduces the need for coax, it puts all the weight in spots. The weight will no longer be distributed between the radio head and the antenna, all in one unit. That makes a difference in these systems. While it's great for system performance, it will change the physical structure of the equipment.

That and it may take more power to run each active antenna. It will take more fiber to get more data to each antenna. That means the cables will be larger.

The good news is that massive MIMO normally has everything in the antenna, so the radio heads and coax will disappear.

Resources for TIA-222:

- https://www.aglmediagroup.com/revision-h-of-ansitia-222-is-published/
- http://www.safeschoolspg.org/examples-of-cell-tower-fires--collapse--ice-strikes--and-theft.html
- https://natehome.com/regulations-and-standards/standards/
- https://www.tiaonline.org/news-media/press-releases/tia-announces-publication-tia-222-h-standard-antennas-and-supporting
- http://www.americantower.com/corporateus/solutions/towers/wireless/tia-222-h.htm

- https://www.tirap.org/wp-content/uploads/2017/06/PAN_ANSI-TIA-222_May-Jun_2017.pdf
- http://www.towernx.com/downloads/ANSI-TIA-222-H_Changes_Overview_MM_v2_12-15-2016.pdf

Summary

I hope this has helped you realize that changes coming up will add even more work to the tower industry. Mounts and the labor to replace them will be needed. Of course, site acquisition wins big time, but then the changes need to be made. The workforce will be overwhelmed in 2019, and this is only going to add to the work.

One more thing, many cities, townships, and municipalities are limiting the building of new towers. I have seen the local governments make a lot of money off the permitting of new equipment on the tower; now, they must make sure that the process is followed. While they don't want new towers, the requirements for existing towers is increasing. Most times it would make sense to build a new tower instead of trying to load the existing, but if the township says, "NO new towers," then you need to improve the existing structure to hold more crap.

We also need to make sure they don't collapse, fall over, bend in half, and that the antennas don't' fall or blow off the tower. After all, someone could get hurt. Let's be as safe as possible on the tower, AND let's make the tower as safe as possible for the people below. Rev H addresses the issues that caused problems in the past. It keeps the professionals on the tower.

Wireless deployment is not cheap; it takes time, planning, and money. Lots of money. Site acquisition costs money; planning costs money, equipment costs money, along with installation, design, testing, and so on. It all adds up. That's why the professionals do it right, they follow the rules, and they know what needs to be done.

One thing that the MIMO system should do for the carriers is that the system should last longer than 3G or 4G, in fact, the changes the carriers make from here on out will be through software if they have enough spectrum. That is why the carriers need to plan this carefully. Sprint is in the driver's seat; they have over 100MHz of spectrum in the 2.5GHz TDD band. It is ideal for 5G, and it should allow them to deploy MIMO and use that spectrum for a long time to come. That is if they plan accordingly.

Tower Crew Summary:

By the time the tower crew gets to the tower to swap or install this equipment, there should be a MOP, Method of Procedure, already in place. We all know that this is a good guideline, but each installation has its nuances.

First off, know the scope of work, (SOW), before you go. Take the time to know what your installing, which OEM it belongs to, and how you intend to rig the equipment. Many of these antennas are heavy, and they may not have an easy way to attach a rope to the antenna. This is going to be a challenge, and each OEM has slightly different dimensions.

Second, look over the SOW and the MOP, make sure you know what you're up against.

Third, know the tower. Do the site walk, look for problems. Look for anything that could be an issue. If you were trained properly, you will walk the site and look for hazards on the ground and in the air. It should be part of your training.

If you're removing equipment, then take the time to look for issues that could slow you down. Make sure the mount is what they say it is. Just because it's documented somewhere doesn't mean the documentation is correct. You may need to replace the mount if it doesn't match the paperwork. The mast may be undersized, or the radio heads might be hard to reach. Pay attention to detail.

Any way you look at this, it's dangerous work. Safety is obviously the number one issue. You need to know what hazards are at the site. The best way to do that is the initial survey and toolbox talk. Experience helps. After a few installations, your awareness of what to look for will be there. You know how to rig in most cases. Experience helps.

Don't forget about the cable. It's big and bulky. It will be a challenge to run up the tower. Think about that when doing a site walk.

Resources:

- https://pdfs.semanticscholar.org/a33b/254b477253d6342bf9c54835ec763e1695af.pdf
- https://networks.nokia.com/solutions/massive-mimo
- https://wade4wireless.com/2017/11/27/what-is-massive-mimo/

- https://wade4wireless.com/2018/01/29/about-massive-mimo-beamforming/
- https://wade4wireless.com/2018/03/12/size-matters-with-massive-mimo/

What does it mean for the suppliers and GCs?

Here's the deal, the equipment we have today will not cut it. Replacing the existing BBU and radio heads and antennas with the new system. It must happen. So, this means a lot more work for the deployment crews. As always, it will all happen at once, causing a strain on the tower workforce.

The carriers need to invest in this, apparently, the stepping stone to real 5G bandwidth and performance. Enough said. They will see the benefits in system performance and system loading. It will give them a reason to say they are 5G ready.

Who benefits?

- Tower crews because they will be busy replacing equipment on the tower and the ground. This is going to be a step towards 5G in more than just software. This is a hardware change that will require work to be done at the site. It is going to take new antennas, new radio heads, and new BBU equipment. This is good for the tower crews to make the most of what is available. They will get work making site improvements. Tower and ground. It must be improved, upgraded, and new. The key to this is to make it as 5G ready as possible, but that is the job of the OEMs to make sure that future work will be done remotely and efficiently.
- Engineering teams may benefit because new RF engineering and optimization along with drive tests to complete the rollout of the new system. Let's not forget the potential for self-interference that's going to happen. There is a learning curve.
- Backhaul providers because:
    - The fiber needs to be upgraded, new fiber or additional fiber.
    - Wireless backhaul will need to be replaced, added, or upgraded.
    - The router needs to be upgraded or replaced.
    - Backhaul will need to be upgraded, so the service provider had a real opportunity to make some additional money on installation and possibly monthly reoccurring for fiber delivery services.
- The OEMs will have a significant push for getting the new gear out to carriers, mostly in urban areas, so it will not be the entire system that gets upgraded, but the specific markets where loading is needed or where 5G is a priority. The good news is that it will be new hardware

and software to make this happen. While the carriers will not like this because it's a major upgrade. New antennas, radio heads, and probably BBUs. All that equipment at the site must be upgraded. The OEMs will have to make this equipment 5G ready. Not just 5G ready, but ready in such a way that no more tower work will need to be done. It is going to be a great undergoing, but one that needs to be done to hit the bandwidth of 5G proportions. While in the beginning, this will be LTE on steroids! That's right; high bandwidth is the predecessor to true 5G. It's more than just a simple upgrade and chances are good they will be done in the urban markets. First, that is where they need the bandwidth. The OEMs have a shot at selling hardware, software, and possible services to the carriers to make this happen.

- Fiber vendors will see an uptick in business because the fiber for the fronthaul and backhaul will go it and need to be expanded and improved. Fiber will increase. I am curious if the carriers will continue buying the hybrid cables for fronthaul or if they will just buy armored fiber lines to run to the radio heads up the tower or in the rooftops. They still need power to the antenna, which they initially needed for the radio heads. Remember that the connectors may change, so we will have to think about the distribution of jumpers.
- Tower mounts may change. All the weight from the radio heads will be shifted to the actual antenna mount. Whether it is a simple mast of the whole mount. The radio head weight will be gone from behind the antenna, and now it is going to be in the antenna. Weight distribution has changed from being evenly distributed to being concentrated on the antenna. Site engineering should be fun.
- Possibly power upgrades will be needed which means potentially new rectifiers and battery upgrades and then utility upgrades. Remember that we were trying to get more power efficiency at the sites, this may be a setback for that effort. If you need a new rectifier, then maybe new or additional batteries and associated cables and hardware. Then the power from the utility to power said rectifiers.
- Routers need to be upgraded or replaced to handle the increased bandwidth needed at the sites. Possibly the distribution of bandwidth at the site may require fiber over CAT5 or CAT6. These are all changes that may need to be made at every site the carriers add massive MIMO.

Think about the changes at the site, it all takes more and more bandwidth which is going to be in the Gbps range now, not the Mbps.
- Vendors who sell Hybrid cables. The hybrid cables will need to be upgraded or replaced at the sites. These are the cables that have both power and fiber in the same cable that is run up the tower. So, this cable is normally armored and weatherproof with everything inside of it. These cables are specially made and will be made for the rollouts then stocked somewhere. They are very heavy and special purpose riser cables. They will probably have a run based off projections, but I see it as a run that will last 2 or 3 years, maybe for each carrier will make up to 15,000 cables, maybe more depending upon where they intend to add massive MIMO.

Who doesn't benefit?

- The tower companies may not get any more money. The way most leases are written leaves room for these upgrades. Of course, the tower companies will find ways to make incremental dollars like site access, new structural for less weight, and so on. What they will not get is the additional rent money that is their bread and butter. However, we will see. Those larger tower owners are savvy. They will do all that they can to make sure they get something, but if the form is smaller, I am not sure how they will get more money. Something else to consider is that these are mostly in urban areas meaning that they may be mostly rooftops.
- What about small cells? I bring this up because, in theory, the way that the massive MIMO improves densification it may reduce the need for a small cell that usually would fall within the coverage area of the tower for loading purposes. This hurts me to say, but small cells will be a fill site outdoors. Indoors we will still need small cells for coverage and offloading. The outside in coverage does not cover as well with new environmentally friendly windows. We need the indoor small cells more than ever.
- Coax vendors won't see much of an uptake. There will be no more coax at the tower sites, less copper and more fiber. Looking at the additional parts associated with the deployments we need to see that coax is going to be reduced dramatically. Coax, hardline, jumpers, and associated connectors and ground kits will no longer be needed for this type of

deployment. Get ready to see a lot more of the old stuff on the scrap metal places.

- CAT5 jumpers have traditionally been used everywhere. While they will be using some CAT4 jumpers, fiber will become the new normal for jumpers. You may not think this is a big deal, but there are a lot of CAT 5 jumpers used at every site. Now they will start using more and more fiber jumpers.
- RF Design teams, it seems like most carriers will use programs or internal teams to design the system. Much of this work has been moving to India, but the carrier will use their existing data to build models. The good news is that now they should be using 3D data to calculate the coverage in buildings and high rises along with standard ground data.

Economies of size with Massive MIMO

Size matters! You heard me, size matters! Why, because size and weight are what the tower companies will be looking at for the new massive MIMO antennas. Let's refer to the massive MIMO antennas as active antennas

Overview

Massive MIMO antennas are active antennas because they have both the radiating elements, like all antennas, but they also have the radio heads behind them. While the designs are slightly different between TDD and FDD, the idea is the same. A dedicated radio head for each element in the antenna.

In this analytical report, we look at how the massive MIMO active antenna will be different from a normal antenna with radio heads attached and how that will affect the payback of installing a new antenna. When looking at new technology, we must analyze the payback. There are several factors that could make the new technology too expensive to deploy to not practical in the real world.

I focus on size in this report but also on why the sizes are different in each spectrum and technologies. It is not a good idea to take this and make massive MIMO black and white. We have to go deeper to look at why some bands are more attractive than others, or why TDD might be more attractive than FDD in massive MIMO and moving into 5G. There's more to it than that. Maybe we have to look at what is better for our system, like a 32x32 array over 128x128. We'll dive into that in the report.

For now, I wanted to go over why this is a matter that the carriers will have to weigh out when deploying massive MIMO and why some carriers may have decided to go with fixed wireless first instead of rolling out mobile massive MIMO.

OK, that's enough to get you started, let's get to it and don't be afraid to give me feedback at wade@techfecta.com so I can tailor these reports to your liking.

Why does size matter?

Let's look at it this way; bigger antennas cost more money all the way around. Size has a direct correlation to costs. Let me break down what costs more with a bigger antenna. When looking at size, remember that the bigger the antenna, the heavier the antenna and that also plays a part here. Keep that in mind because I cover it farther down.

- Larger antennas:
    o They cost more to build and ship. The OEM will charge more. CapEx.
    o The installer will charge more to install the unit. CapEx.
    o The tower company will charge more to put on their towers. OpEx.

The point here is that there must be a balance. The carriers know that payback needs to balance out with the costs. That's where we find balance, between the costs, CapEx and OpEx, and the payback, number of subscribers and improved performance. The balance is subtle but important.

These active antennas may not make sense to put everywhere. Do we really need to put them near a farm where there could be a total of 20 users at any given time? Probably not unless one of those users is a CEO or a president. Power and position have privilege.

We'll look at what effects the size.

Size is key to much of what the carrier pays for, along with weight. Let's look at what factors change the size of the massive MIMO antenna.

- Frequency matters. I'll make this simple, the lower the frequency, the larger the antenna. It's that simple.
- TDD or FDD matter because with FDD you will have 2 sets of radio heads and TDD only has one. FDD will be bigger because 2 sets are larger than one. I'll explain more below.
- Size of massive MIMO, meaning the number of elements. If you have 32T by 32R, (32x32), you have 32 transmit and 32 receive elements. It doubles each time, 64x64 has 64 of each element and radio head, 128x128 has 128, and so on. The more elements and radio heads, the larger and more expensive the antenna.

What about weight?

The larger the antenna, the more it weighs, so that is obvious, but I thought I would point it out. Weight plays a part the same as size.

- Heavy antennas cost more because:
    - Installers must mount them, antenna or rooftop; it will take longer if the antenna is 300 lbs. instead of 100 lbs.
    - Towers need to be able to handle the excessive weight structurally, so the carriers may have to upgrade the towers if the massive MIMO antenna is heavier than all the equipment that comes down, this is a CapEx charge.
    - Tower companies will want more rent money if the antenna weighs too much, this would be OpEx.
    - OEMs costs go up, CapEx.
    - Shipping costs go up, CapEx.
    - Maintenance is an issue if a massive MIMO antenna needs to be replaced. Probably OpEx.
- What would add weight to the units?
    - Remember that these are active antennas, so they require power, they have active electronics in them. This adds weight.
    - They have small radio heads in them. Each radio head is connected to an element. The more radio heads and antennas that you have the heavier the antenna will be.
    - Larger elements in the lower spectrum will add weight.
    - Larger antennas require heavy duty mounting hardware that adds more weight.

Is there a difference between TDD and FDD?

What is FDD?

- **FDD** – Frequency Division Duplex is something that was used commonly in 3G. It's paired spectrum with an uplink band and a downlink band in their specific spectrum. For 1G, 2G, and 3G this was common, so you could have a talk and receive channel in the system. TDD was valuable for voice in CDMA, it made things simple and worked well. There is a guard band in between the transmit band and the receive band. FDD was very popular with GSM and CDMA. Now, with VoLTE using all data, it doesn't seem to be as valuable as it once was.

What is TDD?

- **TDD** – Time Division Duplex is where there is one large piece of spectrum used for uplink or downlink. Any part or percentage can be assigned to be the uplink or downlink. If you have 20MHz of bandwidth available, then you're not locked into 10MHz up and 10MHz down like FDD. Instead, you have full control over how much goes up and comes down. However, Wi-Fi spectrum is pretty much all TDD, and it works quite well for data. On the other hand, WiMAX used TDD, and it never fully blossomed until eventually it was cast aside for LTE. TDD makes MIMO technology easier to use because it is all in one band.

Learn more at https://wade4wireless.com/2017/01/16/an-overview-on-tdd-and-fdd-formats/.

Yes, the format of the carrier has a serious impact on the size of a massive MIMO antenna. First, remember CDMA? When we had CDMA, FDD was all the rage. To have the dedicated spectrum for uplink and downlink made all the sense in the world. Then, with LTE, we thought it was nice to have dedicated spectrum each way, but the reality was that it became less of an issue with carrier aggregation, dynamic uplink and downlink balancing and VoLTE. Hey, maybe Wi-Fi had it right all along. LTE is benefitting from Wi-Fi's "lessons learned." Just like the massive MIMO technology, Wi-Fi had it first. LTE is putting that technology on steroids. Then 5G NR will amp it up even more. How cool is this?

If you think it doesn't make a difference, it does. You see the carriers loved FDD in the CDMA world because they had efficiencies when the uplink and downlink. FDD allowed them to have dedicated uplink spectrum and downlink spectrum. This was a crucial factor for voice efficiency. Even with LTE is seemed to be a good thing when they had the spectrum broken apart.

That was until LTE had VoLTE for voice. VoLTE is Voice over LTE, which is a real digital voice. The efficiency of uplink and downlink balancing was not possible when dedicated spectrum up or down created more problems than it solves. Sprint's 2.5GHz spectrum and the CBRS 3.5GHz spectrum looks quite sexy. It allows the carrier to control uplink and downlink dynamically free of any dedicated up/down spectrum barriers. Awesome!

For instance, LTE is more like Wi-Fi now. It can be more efficient when you can have the spectrum controlled by the carrier, not dedicated. I wondered if the carriers would think about trying to change their new spectrum. It seems like now, FDD having dedicate spectrum would create limitations. Wouldn't it be nice to control what goes up and what comes down in LTE and especially in 5G? TDD allows that because it is transmitting, Tx, and receiving, Rx, on the same elements and in the same spectrum. How cool is that? Just like Wi-Fi, only it's LTE, soon to be a 5G format. I would think 5G will be LTE on steroids.

Why does this matter in massive MIMO? Again, the FDD system will need dedicated antenna elements paired with radio heads for transmitting and dedicated elements and radio heads for receive. Therefore, a 32x32 active antenna would have 32 transmit and 32 receive elements paired with a radio head port in the antenna which would effectively look like, in my mind, 64 heads in one antenna.

A TDD system could have the receive and transmit together on one element. Therefore, a 64x64 active antenna would have just 64 elements paired with radio heads.

At least this is what it's looking like right now. So, half the number of elements for twice the performance, in theory.

The antenna that has half the elements should be half the size, smaller antennas with less weight make for a happier installation, lower costs, and more effective rollout.

Beck to cost, elements and tiny radio heads all cost money. The payback and gain by adding more active elements have to have balance somewhere. If 64x64 costs 5 times as much as 32x32 it may not be worth putting it in. If 128x128 costs 10 times as much, then when is the payback? There has to be a balance between antenna cost and system gain.

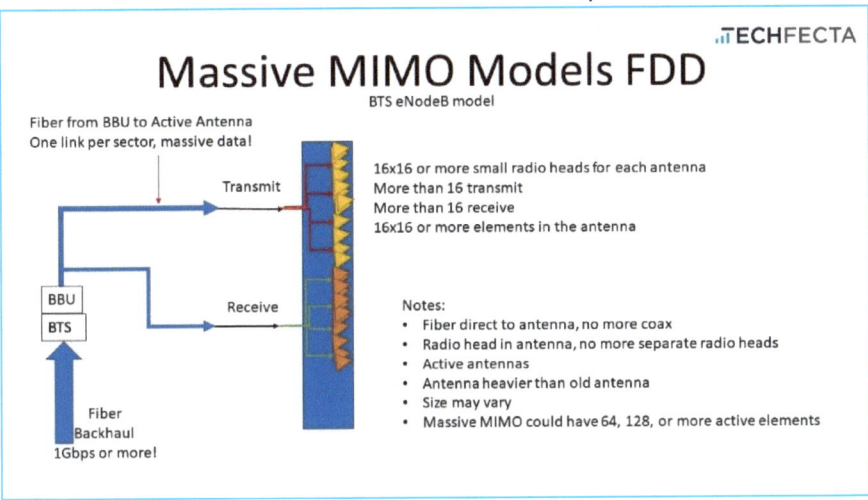

Figure 9 Massive MIMO FDD Model

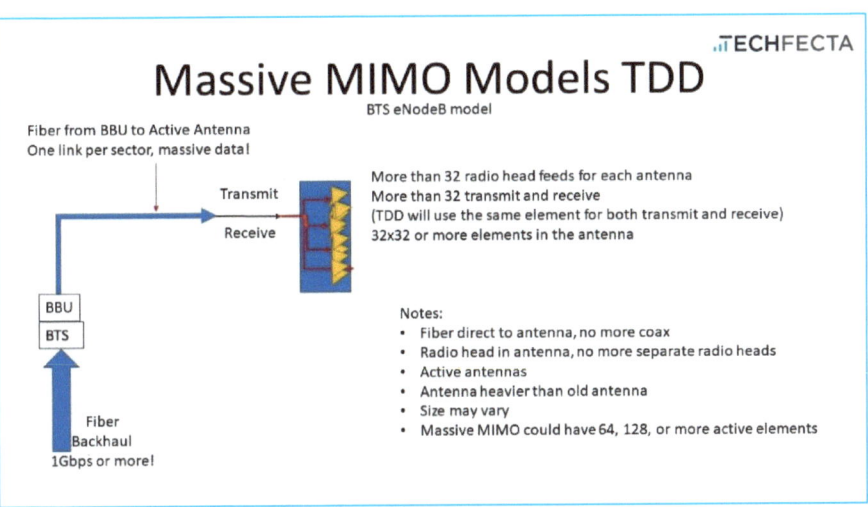

Figure 10 Massive MIMO TDD Model

What about frequency?

How does this affect the antenna? Well, the antenna size is determined by the band. The lower the frequency, the larger the antenna, or at least the elements. That's a normal antenna. Now that we have massive MIMO, it makes more of a difference because the radio heads are behind each element in the antenna. This can be a factor in antenna size.

The lower bands like 1.3Ghz and lower, are going to have larger antennas that require more size just due to the lower spectrum. That is if they want 3dB of gain or more. There are many factors with antenna design which I am not going to get into, but the lower the spectrum, the larger the antenna. Remember that the carriers want plenty of gains and need to have the efficiency to put the least number of antennas on a tower, say 3, as possible. If it is a mini macro on a pole or a small cell, then you may rely more on one or two antennas to cover what you need. Lower spectrum makes that more of a challenge.

While you think it may not matter, you do not see the bigger picture. Larger antennas cost money, and many carriers have spectrum in many bands. In fact, why do you think that T-Mobile wants the CBRS 3.5GHz spectrum so badly? They see the value in the short-range coverage. It's high spectrum, smaller radios, and antennas, and covers the smaller areas efficiently. The deal with Sprint fell through, now they need a contingency plan, and the CBRS looks inviting.

How much is too much?

Here we have the real conundrum of massive MIMO. How much is too much? Do we know the payback of massive MIMO? It looks like we need it for true 5G to roll out with all the promise we expect of 5G. I mean it's more than just the new format of 5GNR, it's all the features that give us Ultra Reliable Low Latency, URLL, and extreme broadband.

There has to be a balance of where we put it, how we deploy, and so on. It makes sense to put it in an urban area where the payback is immediate. Lots of users can justify the cost. If we are covering cows on an IOT system, then it doesn't make sense, does it?

If the cost of a 64x64 is 1/3 the price of a 128x128, then it may make sense to go with the 64x64 for the payback. The number of radio heads will change the price of the unit along with the size and weight. We have to be financially responsible, don't we?

Larger antennas cost more.

Then, there are the mounting issues. They will leave it up to the construction crews to install the equipment, but they won't like putting the monstrous active antenna on the towers if the tower companies raise the rent 10 times. They also have to consider the tower modification implications. There has to be a balance.

Now, for someone with a TDD system if they find the right model. If the model makes sense, then they could lighten the load on the tower. This may or may not make the tower companies happy, they want more rent, but they don't want to modify the towers if they don't have to. Actually, they pass that cost onto the carrier, so maybe they don't care.

For the FDD systems, they will have to install larger active antennas because the Tx and the Rx will be split. You need 2 active element arrays. This added size, cost, and complexity of the system. However, it will enhance the performance of the system. You no longer need radio heads and coax jumpers since it is an active antenna.

How has this changed from the traditional models?

That depends on which traditional model you're looking at. If you look at 1G and 2G, they had the radios on the ground and ran coax cable up the tower. They were relying on more power to reach the masses and were pretty much voice only.

If you look at 3G, you had a digital system where the carriers started to put the radio heads up on the tower at the antenna. They would run fiber and copper up the tower but still put many of their radios on the ground.

When 4G came along, and the higher spectrum had to be used, then it was definitely fiber and copper for data and power to the radio heads then coax jumpers to the radio head.

Now, with massive MIMO and 5G, we will see the rise of active antennas where we still run fiber and power to the top of the tower, but to the antenna itself. No more radio heads or coax will be needed for these systems. Now we will start relying on higher and higher spectrum. Where we once had RF loss through the coax, there is none. All the RF will be in the antenna itself, along with the power supply to power each tiny radio head connected to its own tiny element. There will be hundreds of these in each active antenna for massive MIMO which plays perfectly into 5G NR. Also, 5G NR should be able to put all the spectrum in one band, so the only aggregation you will see is between bands and sites and cells. The advances are amazing when you look back at 2G. WOW! Isn't it exciting that the macro site will be the source of so many features and bandwidth? No more limitations there. It's up the end user to decide what to do next as these systems make coverage available everywhere with massive broadband.

But wait, that's not the big picture!

The reality is, for mobility, we have to look at what we're replacing. If the carriers are going to upgrade to massive MIMO in their existing spectrum and replace their existing equipment, then they have an advantage.

For instance, they will install one unit. The active antenna will have fiber running right to it, direct. There is no longer all the crap on the backend, like the radio head, the coax jumpers, and a separate antenna. All of that equipment adds problems. Let me break it down; the radio heads used to have 1 to 3 fiber pairs running to them, that will change, now there will be many more. There is more data, more overhead, and more bandwidth needed. That is why all the fiber will be connectorized.

I know I threw a lot at you, but let's look at everything and what it means.

- No more radio head, less room needed on the tower, the weight of the radio head is probably more than the radio heads in the active antenna. Less weight and one less point of failure.
- No more coax means less weight, no PIM testing, one less point of failure, no reflected power, easier troubleshooting, less time of installation. For those of you that don't know, coax jumpers take a lot of time to make, weatherproof, tighten properly, and secure properly.
- Fiber connectors save a lot of time, in the old day's tower crews had to put connectors on the fiber after they cleaned it and then test it thoroughly, all this takes a lot of time to install.

With everything in one unit, installation is quicker. Mounting should be easier. One unit to install, not many for each sector. However, now we have a huge point of failure, if the active antenna goes, we're down hard for that sector.

One more thing, in theory, we should have electric downtilt with the massive MIMO antenna that will be controlled automatically by the system. Azimuth is important, but now we may not have to worry about the 3 degrees of down tilt like we used to.

Less time to install, easier to install, less equipment hanging on the tower. It's a win-win all the way around. All this with increased performance. WOW!

Pros and Cons:

Pro:

- Fiber to the antenna decreases installation complexity,
- Active antennas are integrated,
- Massive MIMO improves system performance for;
    - Coverage through beamforming,
    - Multi-user, MU-MIMO, allows the beams to talk to multiple users simultaneously,
    - Increased throughput to each user,
    - Increased densification for power and throughput to multiple users,
- No more coax jumpers, PIM testing, weatherproofing, and so on,
- Less weight overall due to less equipment on the tower,

Cons:

- Increase system complexity,
- The increased cost of antenna,
- Could be a single point of failure, not sure about how the connection to the active antenna will work,
- More fiber jumpers up to the tower,
- Probably increase power draw for the active antenna,

Things to think about?

- Cost of the array, does 32x32 serve your needs or can you go 64x64 or 128x128? Which delivers the best cost for the best price?
- If FDD, what size can you put ion the tower? Will it match the antenna size you have now?
- Are you ready to run more fiber up the tower or across the rooftop?
- Will the payback make sense?

How does the massive MIMO system payback the carrier?

- Increase Throughput

- Much better densification, concentrating the power to each UE,
- Better throughput to each UE through beamforming and multiple users talking at the same time, remember that there are multiple radio heads behind each element,
- Less physical complexity on the tower,
- New options to carriers for deployment,
- In urban areas, it could reduce the need for small cells in the macro's coverage umbrella,
- CRAN Massive MIMO greatly improves localized densification,
- Spectral efficiency is greatly improved by the beamforming,

Resources:

- I wrote an article about the differences between TDD and FDD, found here, https://wade4wireless.com/2017/01/16/an-overview-on-tdd-and-fdd-formats/
- If you want to learn more about Massive MIMO, https://wade4wireless.com/2017/11/27/what-is-massive-mimo/
- Learn more about massive MIMO beamforming at https://wade4wireless.com/2018/01/08/what-is-beamforming/ and https://wade4wireless.com/2018/01/29/about-massive-mimo-beamforming/
- The 5G ecosystem, https://wade4wireless.com/2018/01/22/the-5g-ecosystem/
- Basic antenna concepts, https://www.pulseelectronics.com/docs/library/Antenna%20Basic%20Concepts%2007%2012.pdf
- Antenna size and gain, bandwidth, efficiency, https://nvlpubs.nist.gov/nistpubs/jres/64D/jresv64Dn1p1_A1b.pdf
- Antenna basics, http://www.phys.hawaii.edu/~anita/new/papers/militaryHandbook/antennas.pdf
- Antenna design, http://bluflux.com/yes-the-600mhz-white-space-auction-will-impact-rf-design/
- Frequency bands, http://www.antenna-theory.com/basics/freqBands.php

- Why antenna size matters, http://aviatnetworks.com/media/files/Why_Antenna_Size_is_Important_Jul_11.pdf

Spectrum Options

Here is where it gets interesting. The carriers can use what they have now to step it up. However, the one company that could really benefit is Sprint. If only they could get out of their own way. I say that because their history shows repeated mistakes, (merger with Nextel, Xohm, NGN, heavy debt with little to show for it, dropping from 3 to 4 in the US).

It's important to understand that Sprint has the great spectrum to roll this out but not much money. Nor are they structured to make something great happen. Why do I say that? It's hard when you're the discount carrier in the last place. I don't mean to sound harsh, but I rooted for these guys for years, and they repeatedly let me down with their horrible decision making. I would love to see them step up and make this happen in a big way. It would be a boon to the industry. However, if they roll it out will they be able to capitalize on the new asset they have? I hope so. Recent history has shown that they are working to improve what they have. I am hoping that the team of Marcel Claure and Masayoshi Son make great things happen. They seem to be moving the company in a better direction, but it's still floundering in my opinion. I get it, super heavy debt makes it hard to make big moves. That may be why they dropped to number four.

Mobility Connections:

As far as the current business case, the carriers in the USA realize that they are a pipe for all the applications, not much more. That means that the price they get for data is dropping. That is why they are looking to crack into any market they can. The apps are where it's at! What does this mean to the equipment and services provider? It means that costs must be cut. Hardware and services need to get cheaper to maintain that profit that the carrier's investors want to see. While they already have mobile customers, this will improve the quality of service and greatly improve the site loading. So, the experiences will be so much better.

How will Carriers deploy massive MIMO?

I know that Sprint intends to swap out their existing 2.5GHz system for the TDD. They potentially could be the first to roll out a massive MIMO and 5G mobile system. Sprint has the potential to be a game changer if they can get out of their own way. The OEMs are more than willing to support them, but they want to get paid to do the work. Sprint needs to see the value in this. They need to allow the system to be built. However, chances are they will look at the money, not realizing they are micromanaging the system to the point of failure. The company seems to repeat past mistakes repeatedly.

I know that T-Mobile will be rolling out massive MIMO on their new 600MHz system. They too can roll out the first 5G system once the 5G NR radios are ready for nationwide deployment. They know this is not the best spectrum to do massive MIMO, hence, they made the proposal to buy Sprint which can help them move ahead with the 5G way ahead of schedule. The merger will slow them down, but they know once they have control they can dissolve Sprint and take the system to the next level. They will build massive MIMO out and make the 5G portion of it very profitable by leveraging all the marketing they have in place. That is my prediction if the merger goes through, (Written in June of 2018).

Verizon has massive MIMO on the roadmap, but they are not going to move into that realm too quickly on the mobile system. They intend to use it for the fixed wireless when it's ready. They intend to use the mmwave to test it. Why upgrade the old system if they don't have to? They will move into the higher spectrum with the new technology then evaluate if it makes sense to invest.

AT&T has been looking at fixed wireless like Verizon. If they do fixed wireless, then they will likely take the slow and limited approach. AT&T is going to roll out the FirstNet spectrum which could allow them to add massive MIMO at the same time. I think they should, but they may not want to invest any more money into FirstNet than they have to. They already have the business. They may look at Verizon trying to make a public safety play as too much of a threat to go "all-in" on FirstNet. It all depends on what the master plan will be. I don't know what the master plan is, but I do know that they are diversifying into other venture outside of wireless. I think they know that the income there is limited now that everyone is moving to unlimited data plans.

Massive MIMO will be deployed in different ways for different carriers. T-Mobile and Sprint may merge, which will make this an interesting race. I think T-Mobile and Sprint intend to deploy massive MIMO on their mobile systems anyway, so it may not change a thing. It seems like Verizon and AT&T are going to take the more conservative approach of fixed wireless sand only build it where they have to. I get it, safe at first, then aggressive down the road. I only hope they haven't given up on the mobile systems.

New Business Models for the Carriers:

You must understand this is a game changer for the carriers. It can be used in most spectrum out there. Let's look at what advantages we see.

Internet Service Providers

The carriers can finally compete with this technology in a way that they have never competed before. They can align the antennas to serve the users. I'll explain which carriers are in a position to make this happen and in what spectrum, but for now, let's say that the cable companies will finally have some real competition. I am not talking about DISH or DirecTV for TV; I am talking about the ISP access.

The carriers will be able to deliver 100Mbps to the home in a device that is not owned by the cable company. If they can do it for under $100/month, then they stand to put a huge damper on the cable companies because they can offer the end user mobile service and fixed service all on one bill.

TV and Video

The carriers will be able to provide over 100Mbps to end users. This is enough to allow the end user to stop paying the cable companies for TV channels that they don't want and watch what they want on the internet using Netflix, Amazon Prime, HULU or Apple TV. Why pay for what you don't want when you can't watch it when you can watch what you want when you want on demand?

That only makes sense and the cable companies stand to lose even more market share. They have to see it in the urban areas where the carriers will invest in massive MIMO and push to get 5G rolled out as soon as possible. They don't need to invest in TV, that has already been done by the companies I just mentioned. They are the pipe, let the others do the entertaining. Smart!

Today's millennial watches all video on a smaller device, so the winner here moving forward will be the company that can give the user what they want when they want. T-Mobile is already pushing to that demographic. They know that the market wants broadband, so why get it from the dinosaur that tells you what package of channels you have to get when you could order what you really want. The way we view television and video has changed. It's about time that end users have a choice beyond the monopoly of cable for internet and TV.

Massive MIMO and new spectrum will be the accelerators. AT&T and Verizon are betting on the new spectrum like cmwave and mmwave to save the day. T-Mobile seems to think they will deliver what they can any way they can. Who knows what Sprint will do but they have the spectrum to make great things happen, especially with massive MIMO.

IOT

To be honest, I don't see massive MIMO as a mover and a shaker for IOT unless they can use the 2G spectrum. Then they should be ready to move ahead. The benefit of talking to a bunch of devices is a real plus but is the investment worth talking to IOT devices when they will be low-income devices. I just don't see it anytime soon.

The carriers want to be IOT players, that may the way they use their 2G spectrum more than what they have now, but the massive MIMO play will be a part of some of the market using LTE-M, but I think that the bandwidth for IOT is not a factor so much as the low latency and the number of devices. That is where massive MIMO can really make the difference. It can allow one sector to talk to many devices simultaneously. It's just that in most cases IOT is more about a lot of devices that use low bandwidth but need to talk to the controller quickly. Chances are these devices won't be talking on a regular basis because they must conserve battery life.

It seems massive MIMO will have little impact on the majority of IOT, but who knows, maybe there will be a new application that will be a game changer. The great thing about massive MIMO is that is can serve many users at the same time; this is the key to IOT, they will have millions of devices on the system. They don't need the bandwidth and, hypothetically, they will not be talking to the network that often.

Financially there has to be a payback, massive MIMO is expensive, and IOT devices pay very little to be on the network. It doesn't make sense to invest all that money on something that only gives the carriers a few dollars each month.

Don't get me wrong, IOT will take off, but it will be scattered across several industries. Let's save that for another report!

Transportation

Here is an opportunity for the carriers to not only provide bandwidth to vehicles but, in urban environments, they could provide more services to vehicles in highly dense traffic areas. WOW! Massive MIMO will be something that all users would appreciate. They could use it to locate vehicles, control vehicles, track vehicles, provide services, and provide internet access to the vehicle which may have Wi-Fi inside of it.

As the needs of vehicles grown, let's not forget public transportation. Trains and buses will need all the services I mentioned. They will have to take advantage of this. The carriers have a real opportunity to capitalize on this service in cities that would be happy to hand it off for a reasonable cost.

Should 5G be Fixed and Mobile Wireless?

Which will be first to market, deployed, and be 5G? Which system could be the one to bring 5G to the market, for real? Will they deploy fixed broadband to provide service to homes and call that 5G or will they bring the mobile systems up to a 5G standard before the fixed is deployed?

The answer to the title question, "Should 5G be Fixed or Mobile Wireless?" is simple, both. You heard me, both. They need to do both, and the carriers are going to approach this two separate ways. In the USA, some carriers are going to go with fixed wireless. Then they will figure a way to parlay it onto mobile units. The other is going mobile first then figure a way to sell to fixed customers.

There you go, both results will eventually be the same, it's just a different approach. In the end, they both want to serve the same customers; it's a different approach.

How does this tie into massive MIMO?

How does this tie into massive MIMO? Because both systems will employ massive MIMO to serve the customer, they need to serve. No matter what the system is, they need to pass the same amount of bandwidth to as many customers simultaneously. This is done with massive MIMO.

In the US most of the fixed wireless will be in the 28GHz spectrum. Most of the carriers have this spectrum in the USA. They will also rely on the old standby Wi-Fi, but it's hard to charge for something that is normally free. They're hoping CBRS can help; it's another avenue for supporting the delivery of data to the user.

All of these systems will deploy MIMO of some type. Chances are the CBRS will have a limited MIMO, and the 28GHz systems will be an all-out massive MIMO system that will serve maybe 10 to 100 fixed wireless customers.

What's the difference?

For those of you that don't know the difference, here it is in a nutshell.

Fixed wireless generally is a fixed link between 2 or more points. It could be a point to point or point to multipoint. The end unit, the user's device, is generally a fixed radio that provides internet connection inside the customer's location. It does not move but stays in one location. Look at it like your cable modem in your home or your cable box, only a wireless connection.

Mobile would be a site that connects to mobile devices, like your smartphone. They are mobile and can be used anywhere there is coverage.

Fixed Wireless Overview

Let's start with fixed, what is the business case? It's to provide broadband to the customers that typically would rely on cable or DSL or someone to provide them with an internet connection. It's now going to be viable to have something in the cmwave or mmwave that could be multiuser and still provide over 100Mbps to a home or small business and small cells out there. Don't forget, we still need fiber to the unit, but now we can take that fiber strand and send it to multiple homes without running fiber to the home, it would be a wireless link from the pole or building to the home or small business. How cool is that? No more wires to the home, other than power, but that could be underground.

This means the business case for broadband to the home, (BBTTH), should, in theory, cost less than running fiber to a home overhead or underground. No more trenching to every home hoping that they will sign up. Just get the fiber to a pole on the street, or a large building then shoots it to the homes.

Sound familiar, well it should. The Wi-Fi companies and the wireless ISPs have been doing this for years. Only now people want more. Most ISPs did what they could to get 1Mbps to home; maybe they could get 10 Mbps. However, in today's world of massive data usage, they want 50 or 60 Mbps to the home. Times have changed, and customers are way more demanding. They won't settle for "good enough," unless they are buying a phone from Sprint, then within 1% is good enough.

So, the fixed business case looks good if it can be appropriately scaled. However, I believe only the carriers can pull this off with success. They have the pockets and the grit to make it happen. They also have the name. Why? I am glad you asked.

There are many wireless ISPs across the USA. They have found ways to help the underserved areas. They have deployed in the license-free spectrum, ISM bands, where Wi-Fi is. This worked to a point. Many of them did not realize what it would take to install on a tower or building, in fact, many of those companies are run by IT people who ventured into the wireless carrier space. They quickly realized it takes deep pockets to maintain crews to do this work.

There are successful companies that deployed Wi-Fi internet access, like Boingo, they have done an outstanding job. They built a model around resorts and airports that work. People are willing to pay for Wi-Fi in those cases. There is a need.

The problem with WISP, Wireless ISP, is that the bandwidth service could be up and down based on weather conditions. License-free spectrum is low power and prone to interference. That is one of the issues they must deal with.

Side note – I worked for a WISP years ago and the business model was not great. The expense was high, and the payback was not what we had hoped. In fact, we made more money from IT services than we did off subscriptions. However, we did have subscribers. I soon left that venture to work with an installation and integration company. We did many installs for WISPs, many who could not pay the bills. It was very frustrating. It's a tough business, especially when people think that license-free spectrum is so valuable. The reality is, it's free for a reason. Low power and cheap equipment make it tough to roll out, although conventional business wisdom tells us differently. You see, the services are still expensive, and if you want to go on a tower owned by a big boy, like American Tower or Crown Castle Inc, you still must pay premium rates. All OpEx expenses that can bleed you dry. I've worked with many companies that tried to figure it out, and many of them failed. Others pivoted into something more reliable. It's not rewarding, and the spectrum in the US is monopolized by the carriers, the deeper the pockets they have, the more spectrum they have.

That brings me back to the fixed wireless spectrum. You can learn more [here](), but the spectrum to be used for this is expected to be in the higher bands, like 24GHz, 39GHz, 60GHz and 70 GHz ranges. Those spectrum ranges are almost entirely LOS, Line of Sight. The carriers are convinced that can change with technology, but I haven't seen it yet.

All the same, look at the feeding frenzy that AT&T and Verizon went on bidding for 28 to 31GHz and 39GHz spectrum. They went crazy to acquire what they could. I would say the licenses will help them deploy across the US to homes everywhere, in theory. They must make the technology work. At least the OEMs must find a way to create proof that it works. I think the carriers already have a business case built.

They already ventured into the FTTH, Fiber to the Home, space and it was expensive. They could not lower the price of deployment as they hoped. They probably thought they could because they drove down tower work so far, but fiber deployment is expensive and tedious. This is all in addition to attaching to someone's home, which they will sue the installer if something is messed up. It's not pretty, but the townships, cities, and everyone else wants a piece of the pie in the order of fees, permits, and other various expenses.

Mobile Overview

Whereas the mobile case is merely upgrading the existing sites to new equipment that is 5G ready. Maybe with all the other features like massive MIMO, carrier aggregation, and so on. All the things that need to be installed bringing broadband everywhere. This should bring the mobile sites up to over 100Mbps. Why waste time on fixed if your mobile carrier can do it and provide you a device to make your home a hotspot? Just do it! If they already have one of your devices, then they may get another one and cancel their cable service or another internet provider at their home. I would! Although, I live in a suburban area so that won't happen for 3 or more years, will it? NO!

However, this is an expectation of 5G, broadband everywhere. While the carriers may not be excited to put even more money into their sites, they have no choice if they want to compete. Your wireless carrier fee is feeding this expansion! The carriers need to deploy all that they can to remain competitive.

The wireless broadband is the way that the millennials get their data. They rely on the carrier for almost everything.

I feel if the mobility broadband happens and they try to use it for fixed, then it may overload the sites. At least with the spectrum that most carriers have. They know this, that's why the big boys want to roll out new systems to support the home internet case. The only exception that I see is Sprint. However, T-Mobile might get creative with their 600MHz spectrum to get it into the homes of the public, if it's enough. It may or may not be. However, if any carrier could do more with less, then they are the real winner. I think they could if they plan it properly, but I'm sure they know better, (at least they think they do).

However, with mobility systems, you could deploy a broadband solution to the home as easy as putting a device in it and setting it up for Wi-Fi. I would think something like Sprint and Airspan's Magic box would be perfect for something like this. It would be easy for anyone to buy it and install it. Just plug it in and see if you have coverage. Awesome and easy, just what any consumer wants without going to all the trouble of fixed wireless.

Why compare fixed to mobile?

I think we need to, so we can better understand which 5G system will be rolling out first. I think the mobile system will be looked at as another upgrade and overhaul of the existing mobile system. Whereas the fixed wireless system could be a new division that brings in new revenue for the carriers. The revenue that standard ISPs and cable companies had before.

When you look at the business models, they are very different. We want to see where 5G will be applied first, in a fixed scenario or on the existing mobile system.

Fixed Pros and Cons

The pros of fixed are that it's a new revenue stream or at least a way to cut the costs of fiber to the home. If they can run the fiber to a pole and connect 5 to 20 houses off one radio, then they saved a whole lot of money in fiber installation, deployment, and permits. Pros are cost savings and new revenue.

Cons are it's new, and it will need to be tested, and chances are there may be problems. They are also running into the cable companies' mainstay. The cable companies have monopolies all over the place, and the carriers need to figure out how to wedge themselves into those markets. It won't be easy. The carrier will invest heavily to do this even without running fiber everywhere.

Pros and cons are it's all new equipment. Why would that be both? New equipment is expensive to deploy and needs to be put on sites. That means new fiber runs, site acquisition, planning, installation and all the expenses that go with it. Even if it's an existing site, all those details must be worked. However, there is no legacy equipment to remove or replace. New system installs are generally clean and easy to work with when there are no customers or just a few customers. Like I said, pro and con.

Mobile Pros and Cons

Mobile will eventually become 5G, but there is more to it than just upgrading the sites. If 5G needs to be a new format other than LTE, let's say a 5G LTE, then the upgrade is going to be costly. The system must work with 3G, 4G, and 5G. All of them. No easy thing. I believe 4G and 5G will not be a problem, but any carrier is holding on to 3G has a significant problem. Let's look at Sprint, I am not aware of VoLTE for mass deployment, so they need 3G CDMA to keep the voice going. Yes, many people still make calls on their smartphones, and I am one of them. This means that the carrier must support all the systems until the migration is complete.

Migration isn't just about the sites. The devices, like smartphones, all must be ready for the new system. Ask T-Mobile how many devices on the street have 600MHz in them. I would guess less than 100. Maybe ask AT&T how many devices have the FirstNet spectrum in them. Again, a meager number.

The site work isn't all that has to be thought of for mobility; the UE devices need to be ready for the new service.

So, the pro is there will be more bandwidth at existing sites, new features, and bragging rights. All the carriers want to have 5G running on their system just to say they have 5G running on their system. I want to say that, and I don't have a system.

The con is that the equipment at the site must be upgraded. Chances are these are all live sites, would be service affecting to customers. Not an easy thing. It may be day work or maintenance window work. Either way, chances of a live site going down for some maintenance are 100%. Chances are good the migration could be done in steps, and I see massive MIMO being deployed. That means that the antenna and RRH will be replaced with an active antenna. Good and bad. Good because the form factors and weight will be less, along with fewer coax connections. The bad is that all the leases will need to be amended, tower work must be done, and CapEx goes up for a few years during deployment.

Another con is the UE devices will need to be sold to customers. There may be a boost when it first comes out, but the legacy users will hold on, and it will be a long time before they can sunset old products. To see the results, users must have new devices.

Pros and cons are the backhaul and fronthaul. Carriers will need more fiber at the site. Carriers will need new routers to handle the amped-up broadband. Guess what; more backhaul bandwidth means that the fiber provider may need to light up more strands. While this sounds awesome, more bandwidth, for the carrier it's more OpEx expense, meaning that monthly costs go up at every site. Imagine if you have 15,000 sites and the monthly cost for backhaul alone goes up to $1,000 each month. That's $15 MILLION dollars each month, which adds up to $180 MILLION dollars a year, for the rest of that sites life. That's going to be hard to pay for with unlimited data plans.

Who wins?

Up front fixed will claim 5G first but mobility always wins because the devices are already in the hand of mass users. Working devices can see results immediately, even if it's 4G LTE, if people see 100Mbps of throughput, then it's close enough to be called 5G, even though it's not. People want to see results immediately.

However, in the long run, both models win because the revenue streams will continue to increase for all the systems. The fixed will be new revenue competing against the cable companies. Let's go deeper than the carriers. In the fixed arena AT&T and Verizon have the edge spectrum and a plan. They already are testing. They already secured spectrum. They will win the race there.

Cable companies will be hurt by this new push, the way I see it. I am not sure what their defense will be, but I am sure they will think of something.

In mobility, T-Mobile is already pushing to win the broadband race. I would love to say Sprint has a chance because they have so much spectrum, but can they spend the money to make it happen? I don't know.

I'll tell you this, no matter which system is deployed, the fiber and router companies win. The new bandwidth demands require a lot of bandwidth. So, the FTTP, Fiber to the Premise, suppliers like Zayo and ExteNet will be winners. Fiber deployment teams also win. Fiber providers are the real winners though; they will get more money for the fiber that is out there. It will be a big win for them for years to come as broadband needs increase, or at least maintain.

Deployment teams will get plenty of business for the next 3 or 4 years. All the carriers want to deploy. They will all do design, testing, and integrations. It all must be deployed.

Asset owners should get a lot of business, but let's clarify. The equipment on the tower will get smaller and lighter. There should be less equipment on the towers and rooftops. It doesn't mean it can't do more; it just means that it is in a smaller package.

The site acquisition teams will also get a lot of work, no surprise, they are needed at every turn for the permitting, the zoning, the planning, the lease amendments. The carriers try to bring this in-house, but they still need feet on the street in the local markets thanks to all the permitting requirements.

Summary:

The carriers are looking for new revenue streams. I think that is why AT&T and Verizon paid billions for the 24and 28 GHz spectrum. I think they know they must break into new markets as cost-effectively as possible to build a new market up. If they already have this spectrum, that's something to work with. It's all good! It's one more market they think they can tap.

The mobile market is not yet saturated. They are looking for new revenue there, such as cloud services and IOT services. That is all based on quantity, meaning they need a lot of devices to make some money.

AT&T is going to rely on the FirstNet business to happen and bring in some government money that Verizon had tapped for so long. However, remember that FirstNet participation is voluntary, so Verizon may be able to keep most of its customers. That's another article.

In the end, the carriers will find a way to put the fixed wireless spectrum in the mobile devices to add to the coverage. Most people are using the mobile device for data when sitting somewhere anyway. While the usage on trains and buses is key to success, most people are using the systems in a restaurant or park or in an office. Let's face it; if the carriers can offload to the fixed wireless spectrum when the systems are not so busy, they will do it in a heartbeat!

If you want to learn more:

- Fixed Wireless is a Focal Point of 5G

- [5G Fixed Wireless Spectrum and Why it Matters](#)
- [Fixed Wireless Access Overview](#)
- [Learning 5G in the Real World](#)
- [The 5G Deployment Plan Handbook](#)

MIMO Report Summary:

By the time the tower crew gets to the tower to swap or install this equipment, there should be a MOP, Method of Procedure, already in place. We all know that this is a good guideline, but each installation has its nuances.

First off, know the scope of work, (SOW), before you go. Take the time to know what your installing, which OEM it belongs to, and how you intend to rig the equipment. Many of these antennas are heavy, and they may not have an easy way to attach a rope to the antenna. This is going to be a challenge, and each OEM has slightly different dimensions.

Second, look over the SOW and the MOP, make sure you know what you're up against.

Third, know the tower. Do the site walk, look for problems. Look for anything that could be an issue. If you were trained properly, you will walk the site and look for hazards on the ground and in the air. It should be part of your training.

If you're removing equipment, then take the time to look for issues that could slow you down. Make sure the mount is what they say it is. Just because it's documented somewhere doesn't mean the documentation is correct. You may need to replace the mount if it doesn't match the paperwork. The mast may be undersized, or the radio heads might be hard to reach. Pay attention to detail.

Any way you look at this, it's dangerous work. Safety is obviously the number one issue. You need to know what hazards are at the site. The best way to do that is the initial survey and toolbox talk. Experience helps. After a few installations, your awareness of what to look for will be there. You know how to rig in most cases. Experience helps.

Don't forget about the cable. It's big and bulky. It will be a challenge to run up the tower. Think about that when doing a site walk.

More Resources:

- https://pdfs.semanticscholar.org/a33b/254b477253d6342bf9c54835ec763e1695af.pdf
- https://networks.nokia.com/solutions/massive-mimo
- https://wade4wireless.com/2017/11/27/what-is-massive-mimo/
- https://wade4wireless.com/2018/01/29/about-massive-mimo-beamforming/
- https://wade4wireless.com/2018/03/12/size-matters-with-massive-mimo/
- http://www.radio-electronics.com/info/cellulartelecomms/lte-long-term-evolution/lte-mimo.php
- http://www.samsung.com/global/business-images/insights/2017/Massive-MIMO-Comes-of-Age-0.pdf
- http://www.radio-electronics.com/info/cellulartelecomms/lte-long-term-evolution/lte-mimo.php
- https://en.wikipedia.org/wiki/Multi-user_MIMO
- http://techgenix.com/mu-mimo-vs-su-mimo-wi-fi/
- https://www.linksys.com/us/r/resource-center/what-is-mu-mimo/
- https://en.wikipedia.org/wiki/Cooperative_MIMO
- http://www.techplayon.com/active-antenna-system-aas-3d-aspect-aas-5g/
- http://www1.huawei.com/en/static/AAS-129092-1-197969.pdf
- https://arxiv.org/pdf/1612.03993.pdf
- https://cdn.rohde-schwarz.com/it/seminario/Massive_MIMO_antenna_OTA_170420_Italy.pdf
- http://www.iracon.org/wp-content/uploads/2016/03/Taro-Eichler.pdf
- https://networks.nokia.com/solutions/4.9g-massive-mimo
- http://www.5gsummit.org/hawaii/docs/slides/Amitava-Ghosh.pdf

- https://www.google.com/url?sa=t&rct=j&q=&esrc=s&source=web&cd=12&ved=0ahUKEwiszf-0o6PYAhVKRN8KHQELCEQQFgh3MAs&url=http%3A%2F%2Fwww.mdpi.com%2F2079-9292%2F6%2F3%2F63%2Fpdf&usg=AOvVaw0F2G4e_0pTONkDQTdRwFm
- https://www.slideshare.net/ahmed_nasser_ahmed/introduction-to-massive-mimo-42252235
- https://spectrum.ieee.org/video/telecom/wireless/5g-bytes-massive-mimo-explained

The Massive MIMO Report

Acronyms and Definitions

- **AI** - Artificial Intelligence or Augments Intelligence.
- **CBRS** – Citizens Broadband Radio Service – in the USA this is 3550MHz to 3700MHz, often referred to as the 2.5GHz spectrum. Learn more at https://www.leverege.com/blogpost/what-is-cbrs-lte-3-5-ghz
    a. **ASA** - Authorized Shared Access
    b. **PAL** - Priority Access Licensed
    c. **LSA** - Licensed Shared Access
    d. **GAA** - General Access User
- **CCI** – Crown Castle Incorporated
- **CLEC** - Competitive Local Exchange Carrier
- **Cmwave** – spectrum in the 3 to 30 GHz range and will most likely be used in 5G for the fixed spectrum but could have mobility potential.
- **CRAN** – Centralized RAN
- **cRAN or C-RAN** – Cloud RAN
- **DAS** – Distributed Antenna Systems
- **FTTH** – Fiber to the Home
- **FTTSC** – Fiber to the Small Cell
- **FTTx** – Fiber to the Anything
- **FWA** – Fixed Wireless Access
- **IOT** – Internet of Things
- **KPI** – Key Performance Indicators
- **LoRaWAN** – Long Range Low Power WAN
- **LOS** – Line of Site
- **LPN** – Low-Power Network
- **LPWAN** – Low Power Wide Area Network
- **LTE** – Long Term Evolution
- **LTE- U** – LTE Unlicensed, generally license-free spectrum in the 2.4GHz and the 5.8GHz ISM bands
- **MEC** – Mobile Edge Computing
- **Metro cell** – larger coverage area small cell
- **Micro cell** – small coverage area small cell
- **MIMO** – Multiple in Multiple out
- **Mini Macro** – a very large small cell, smaller than a macro BTS

- **Mmwave** – spectrum in the 30 to 100GHz range that will be part of 5G and most likely used for fixed wireless.
- **NB-IOT** – Narrowband Internet of Things
- **NLOS** – Near or No Line of Site
- **NR** – New Radio format developed by QUALCOMM to be used in 5G radios
- **OTF** – Off the shelf, a term used to describe common equipment, like servers or routers.
- **PaaS** – Platform as a Service
- **Pico Cell** – small business, lightly load small cell
- **PoE** – Power over Ethernet
- **POTS** - Plain Old Telephone Service
- **PTMP** – Point to MultiPoint
- **PTP** – Point to Point
- **PTT** – Push to Talk
- **RAN** – Radio Access Network
- **ROW** - Right of Way
- **SaaS** – Software as a Service
- **SAS** – Small Cell Antenna Systems, like DAS systems but with small cells only.
- **SCaaS** – Small Cell as a Service
- **SDN** – Software Defined Networking
- **SISO** – Single in Single out (antennas)
- **UE** – User Equipment, like a smartphone
- **URLL** - Ultra Reliable Low Latency.
- **Wi-Fi** – Wireless Fidelity, generally license-free spectrum in the 2.4GHz and the 5.8GHz ISM bands
- **WiMAX** – Worldwide Interoperability for Microwave Access, based on 802.16 set of standards, learn more at https://en.wikipedia.org/wiki/WiMAX
- **VoLTE** – Voice over LTE
- **Femto Cell** – home use small cell

Thank you!

Thank you for your support. I pray that it serves you well.

I want to thank you for your thirsting for technical know-how. The people that want to learn more will grow in every way. I am here to serve that need.

If you need one on one consulting or specific reports, feel free to reach out at wade@techfecta.com or wade4wireless@gmail.com for direct support.

Copyright

First Edition © 2018 by Wade Sarver. All rights reserved. No part of this publication may be reproduced, stored in a retrieval system, or transmitted in any form or by any means, electronic, mechanical, photocopying, recording, scanning, or otherwise, except as permitted under Sections 107 or 108 of the 1976 United States Copyright Act, without the prior written permission of the author.

I am not a lawyer or an actively certified safety expert. This book is completed based on research and my experiences. Safety processes and procedures are constantly updated and improved over time. The material contained is for reference only and may include products, information, or services by third parties. I do not assume responsibility for any third-party material referenced in this book.

This document is a guide to help people and not a guarantee that you will do everything properly. By reading this, you agree that myself and my company is not responsible for the success or failure of your business decisions relating to the information presented in this guide.

www.wade4wireless.com

www.techfecta.com

Cover and design by Wade Sarver

About

Chief Technology Analyst for TechFecta. TechFecta, Tech consulting for the real world. www.techfecta.com

Solutions Consultant, Technology Analyst, Technology Marketer. Author, blogger, podcaster.

Blog and podcast available at www.wade4wireless.com if you want to follow.

Link up with me on LinkedIn, https://www.linkedin.com/in/wadesarver/.

Reach out to Wade on LinkedIn or at Wade@techfecta.com or wade4wireless@gmail.com to send feedback.

Twitter @wade4wireless, https://twitter.com/Wade4Wireless.

More Reports and Books:

The blog at www.Wade4Wireless.com to offer real-world knowledge to wireless workers and investors.

TechFecta is here to help workers, businesses, and investors gain clarity on the direction technology is heading from where it is today.

Other products include:

- The New T-Mobile Consolidation Report
  https://wade4wireless.com/2018/05/13/the-new-t-mobile-consolidation-report/
- The 5G Deployment Plan Book
  https://wade4wireless.com/2017/01/30/the-5g-deployment-plan-book-release/
- The Wireless Deployment Handbook for LTE Small Cells, CRAN, and DAS.
  https://wade4wireless.com/2015/11/12/wireless-deployment-handbook-for-lte-small-cells-and-das/
- Learning 5G in the Real World.
  https://wade4wireless.com/2017/09/18/learning-5g-in-the-real-world/

The Massive MIMO Report

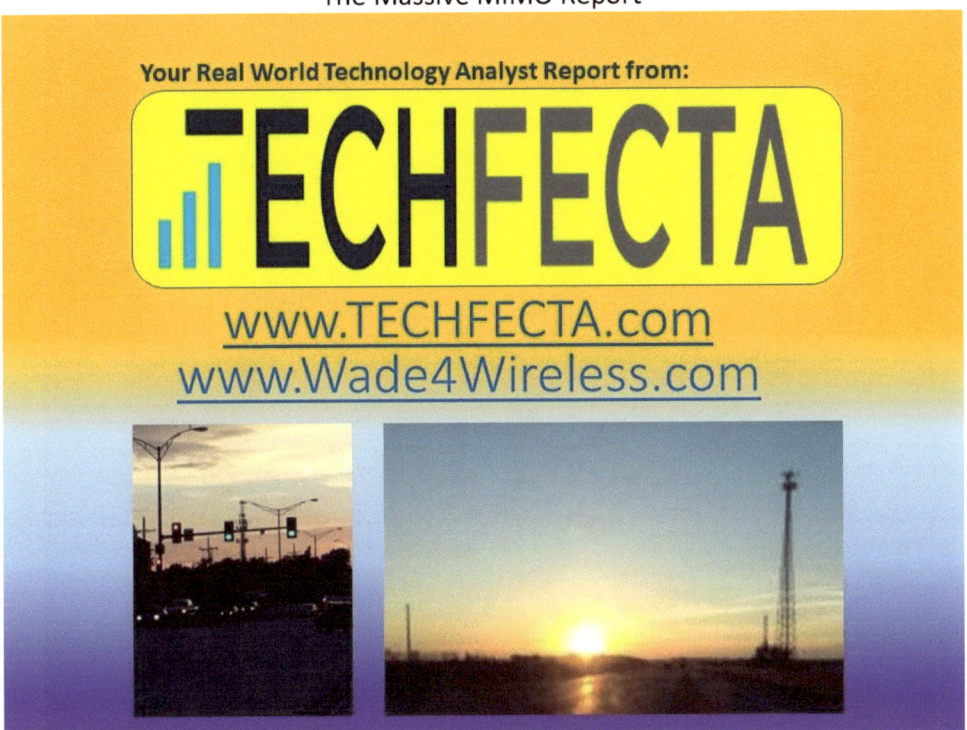

Figure 11 Back Cover